NOMADEN UND SESSHAFTE

Sonderforschungsbereich Differenz und Integration
Wechselwirkungen zwischen nomadischen und sesshaften Lebensformen in Zivilisationen der Alten Welt

Herausgegeben im Auftrag des SFB
von Stefan Leder und Bernhard Streck

BAND 7

WIESBADEN 2007
DR. LUDWIG REICHERT VERLAG

Pastoral Morocco

Globalizing Scapes of Mobility and Insecurity

Edited by
Jörg Gertel and Ingo Breuer

WIESBADEN 2007
DR. LUDWIG REICHERT VERLAG

Gedruckt mit Unterstützung der
Deutschen Forschungsgemeinschaft

Cover Photograph: High Altitude Pasture in the Middle Atlas, Morocco.
(Photo taken by Jörg Gertel in March 2006)

Redaktion und Layout: Ute Terletzki M.A. (Leipzig)

Bibliografische Information der Deutschen Nationalbibliothek
Die Deutsche Nationalbibliothek verzeichnet diese Publikation in der
Deutschen Nationalbibliografie; detaillierte bibliografische Daten sind
im Internet über http://dnb.ddb.de abrufbar.

ISBN: 978-3-89500-552-7
www.reichert-verlag.de

Gedruckt auf säurefreiem Papier
(alterungsbeständig pH7 –, neutral)
Printed in Germany

Contents

Acknowledgements VII

Conceptionalizing Pastoral Morocco

1. Introduction 3
Jörg Gertel & Ingo Breuer (Leipzig)

2. Mobility and Insecurity: The Significance of Resources 11
Jörg Gertel (Leipzig)

3. History of Mobility and Livestock Production in Morocco 31
Jeanne Chiche (Rabat)

4. Neoliberalism, Environmentalism, and Agricultural Restructuring in Dryland Morocco 61
Diana K. Davis (Austin)

Regional Geographies of Pastoral Morocco

5. Social Transformation and Sedentarization in the Eastern Moroccan Steppes 85
Boutayeb Tag (Fes)

6. Pastoralism and Institutional Change in the Oriental 93
Mohamed Mahdi (Meknès)

7. Trading Livestock: Eastern Moroccan Sheep Meat Commodity Chains 107
Mohamed Khalil (Rabat)

8. Marketing from the Margins: The Ilimchan Pastoralists of the Pre-Sahara 117
Ingo Breuer (Leipzig)

9. Animal Production, Herd Mobility, and Rangeland Access in the Middle Atlas 133
Houria Djoudi & Irene Hoffmann & Bouchra El Amiri & Jörg Steinbach (Basel/Rome/Rabat/Gießen)

10. Territorial Restructuring in Agro-Sylvo-Pastoral Systems of Atlantic Morocco 149
Mohamed Aderghal (Mohammedia)

11. Livelihood Security and Mobility in the High Atlas Mountains 165
Ingo Breuer (Leipzig)

12. Pastoral Livelihood Strategies in the Draa and the Souss 181
Jutta Werner (Berlin/Abéché)

13. Social Lines of Conflict between Pastoralism and Agriculture in the Souss 193
Bertram Turner (Halle)

14. Nomads: But how? 211
Hassan Rachik (Casablanca)

Bibliography and Appendices

Bibliography 229

Lists of Maps, Figures and Tables 253

Acknowledgements

Pastoral Morocco examines the history, structure and recent dynamics of pastoral development in Morocco. It assembles knowledge and expertise from various disciplines, such as Geography, Anthropology, Sociology, and Agronomy, and unites scientists from Morocco, Europe and the United States. All contributions stem from intensive, long-term fieldwork and they debate relations between pastoral development, socio-spatial mobility, and livelihood security.

We would like to express our gratitude to those who made this project a success, in particular to the authors. *Pastoral Morocco* is also indebted to the research center Difference and Integration, located at the Universities of Leipzig and Halle-Wittenberg. In its context, the editors conducted five years of research, and organized the international conference 'Globalization and Pastoral Livelihood Security in Morocco', in Rabat in November 2005, under the auspices of the Institute Agronomique et Vétérinaire Hassan II (IAV). Neither the conference nor much of the preceding research would have been possible without Jeanne Chiche, who participated and shaped the project from its very inception. We are equally grateful to our partners from the IAV for the years of productive collaboration. In particular, we thank Fouad Guessous, Mohamed Agbani, Najib Aqesbi, and Fatema Mossadeq. Very special thanks go to Dieter Strauss for hosting the conference in the Goethe Institute of Rabat.

Editing this volume would not have been possible without the tremendous support and efforts of many individuals. Thanks are due to all of them. Most of the papers were translated from French to English, a tremendous task Sandra Calkins performed with outstandingly skillful dedication. Gordon Winder from the University of Auckland carefully reviewed most of the articles. Lea Bauer spent a huge effort in the cartography and map design. Steven Calkins did a final English proofreading. Special thanks go to Ute Terletzki for her dedicated and professional organization in Leipzig and in Rabat, and for her immense effort with technical editing and layout. All of the remaining errors are, of course, our responsibility.

Finally, we are grateful to Stefan Leder and Bernhard Streck for accepting this publication in the series 'Nomaden und Sesshafte'. We also are indebted to the German Research Foundation for financing our research in Morocco, the conference in Rabat and this volume. Last but not least we are grateful to the publisher for producing *Pastoral Morocco*.

Jörg Gertel   Ingo Breuer

Note: Transliteration of terms from Arabic and Berber languages is the responsibility of the individual contributors. Spelling of toponyms that are used more frequently was standardized (e.g. Fes, Marrakech). The boundaries and names represented and the designations used on the maps in this volume do not imply official endorsement or acceptance by the United Nations. The Moroccan currency is Dirham (Dh); in April 2006, 100 Dh amounted to 9.15 €.

Conceptionalizing Pastoral Morocco

–1–

Introduction

Jörg Gertel & Ingo Breuer

Extensive pastoral production takes place on some 25% of the world's land area. It provides 10% of the global meat production, supports between 100 and 200 million households, and comprises herds of nearly a billion head of camel, cattle and smaller livestock.[1] Yet, despite their vital role in food production in marginal environments, pastoral livelihoods are threatened. Rapid urbanization, spreading cultivation, installation of nature conservation areas and hardening international borders increasingly restrict access to pastures and limit spatial mobility. It is however not only the transformations of pastoral territories, but also the exposure of pastoral livelihoods to neo-liberal policies of economic restructuring, changing property rights and newly evolving global marketing chains that generate globalizing scapes of mobility and insecurity. Hence, many pastoral societies are undergoing an unbalanced integration into the global economy, and are subject to manifold processes of specialization, income diversification, and social polarization, ultimately reflected in increasingly unequal access to resources. In this context questions of mobility and livelihood security become central to a scholarly perspective on changing pastoral societies (cf. Gertel, this vol.).

Morocco is integrated and exposed to these dynamics of globalization. In the last two decades the country experienced multiple liberalization programs, the deregulation of economy and trade, and the implementation of a free trade agreement with the United States in January 2006 (cf. Davis, this vol.). Prices of consumer products are increasingly market determined. Remittances from over 2.4 million Moroccans working abroad, predominantly in Europe, and revenues from 2.5 million tourists visiting the country are of crucial economic importance. Nonetheless, Morocco's economic situation still depends on agriculture, which accounts for about 16–23% of the GDP and employs about 40–50% of the country's labor force. Livestock production is widespread; 85% of all farms possess animals and Moroccans own about 16.4 million sheep and 5.4 million goats, many of which are kept in pastoral or agro-pastoral production systems.

The pastoral landscape is characterized by a wide range of natural diversity. The country comprises semi-arid and arid, but also humid regions, and high mountain areas can be found as well as open plains, steppes and desert zones (cf. Maps 1-1, 1-2). Accordingly a large variety of pastoral production systems exists, ranging from vertical transhumance to horizontal steppe pastoralism, and from semi-inten-

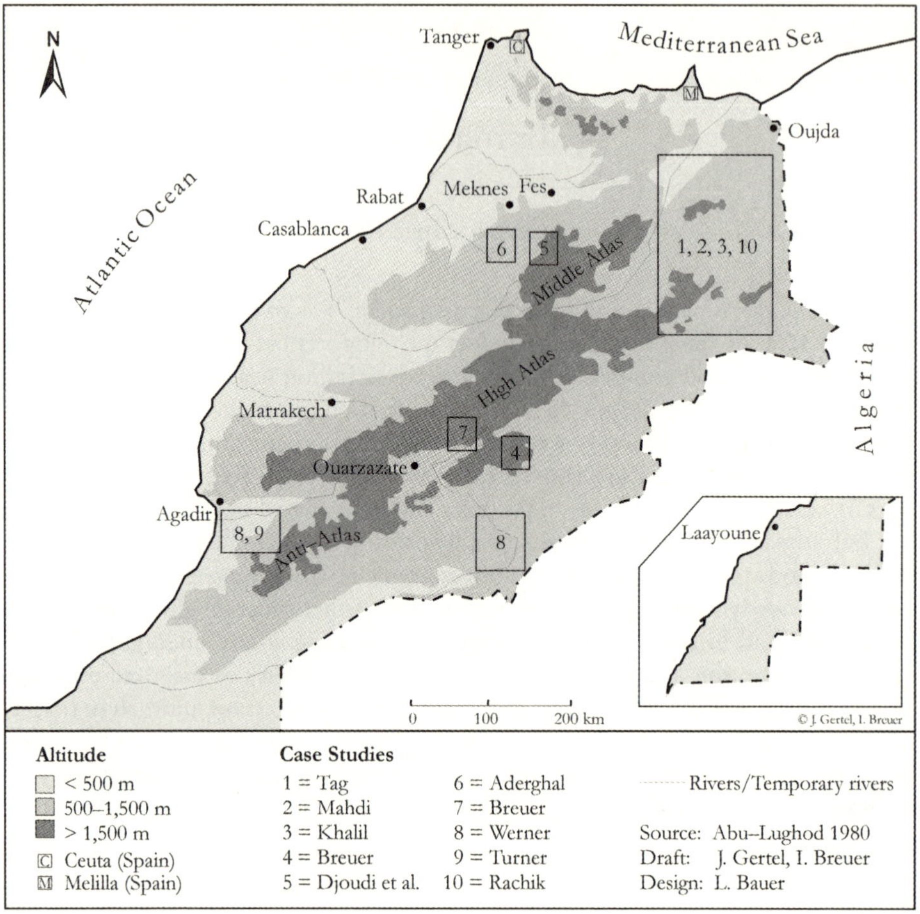

Map 1-1: Case Studies in *Pastoral Morocco*

sive agro-pastoral systems to more extensive forms relying almost entirely on natural pastures.

Large parts of the country are prone to a high variability in precipitation that causes severe inter-annual production fluctuations, exposing Morocco's agricultural and pastoral systems to a wide array of climatic and macro-economic risks. Massive amounts of wheat, the staple food of the poor, have thus to be imported to ensure food security for a growing population of about 30 million, half of which lives in rural areas. While agricultural and pastoral production systems are highly articulated, the agri-food system features deep cleavages between export-oriented sectors, producing, for example, tomatoes and citrus fruits according to EU import quota, and a multitude of small-scale family enterprises that grow food and raise animals for subsistence and local markets. These production systems differ accord-

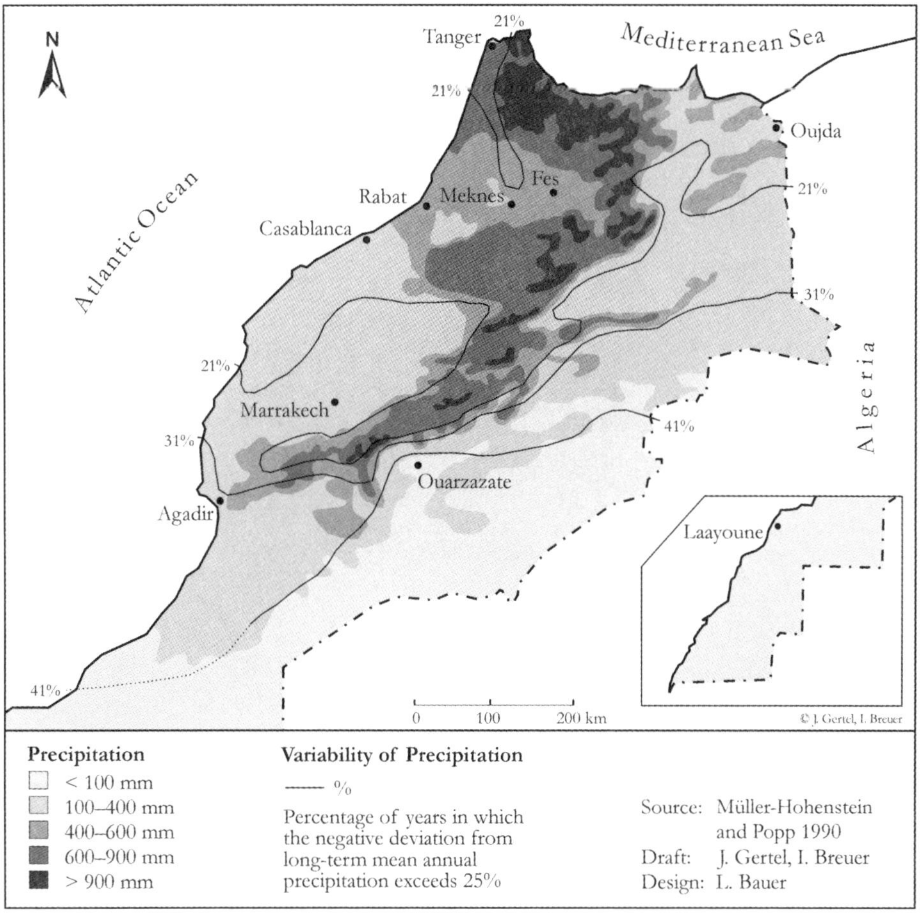

Map 1-2: Morocco – Distribution and Variability of Precipitation

ing to natural diversity, the degree of market integration and state intervention (cf. Chiche, this vol.). The prevalence of poverty in the arid rural regions – precisely the areas that are predominately inhabited by pastoralists[2] – reflect both, structural economic risks and regional specificities (cf. Maps 1-3, 1-4).[3]

Pastoral Morocco offers an exploration into these multiple, and often contradictory processes of globalization and regionalization. It departs from a geographical point of view that focuses on interrelations between pastoral development, socio-spatial mobility, and livelihood security. Four interconnected analytical perspectives are applied. First, pastoral production systems are studied in their relations to market and state interventions, that are reflected, for example, in the commercialization of exchange relations and a closer integration into the world market. Second, pastoral

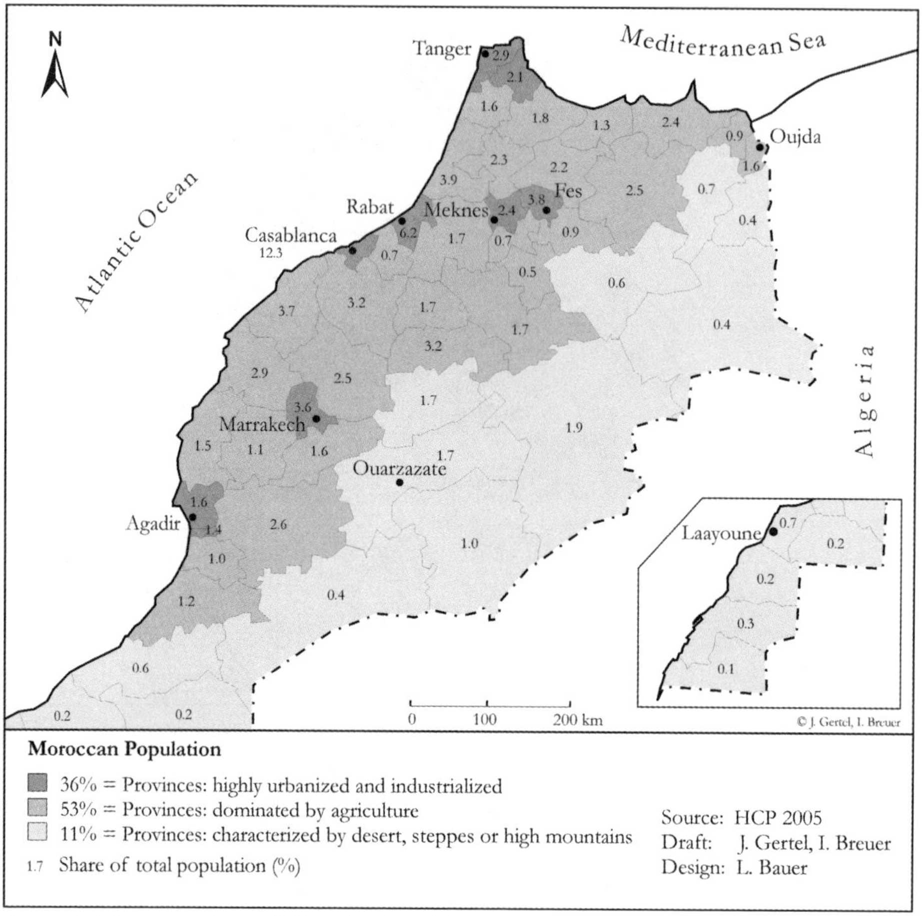

Map 1-3: Morocco – Spatial Distribution of Population

livelihoods are investigated according to new forms of regionalization. These unfold on the one hand when pastoral territories are expanded or contracted – for instance, when boundaries or access rights that control the use of pastures are renegotiated. On the other hand new dynamics unfold due to modern communication and transportation (mobile phones, trucks), creating new social scapes of interaction. Third, as pastoral production is increasingly intertwined with labor migration, different forms of mobility are analyzed and comprehended as strategies to maintain and enhance access to resources. Finally, the intersection of these processes in time and space is analyzed, to examine how new forms of security and insecurity are generated and inscribed in pastoral livelihoods. Here the case studies offer specific insights into the social logic of Morocco's pastoral production systems.

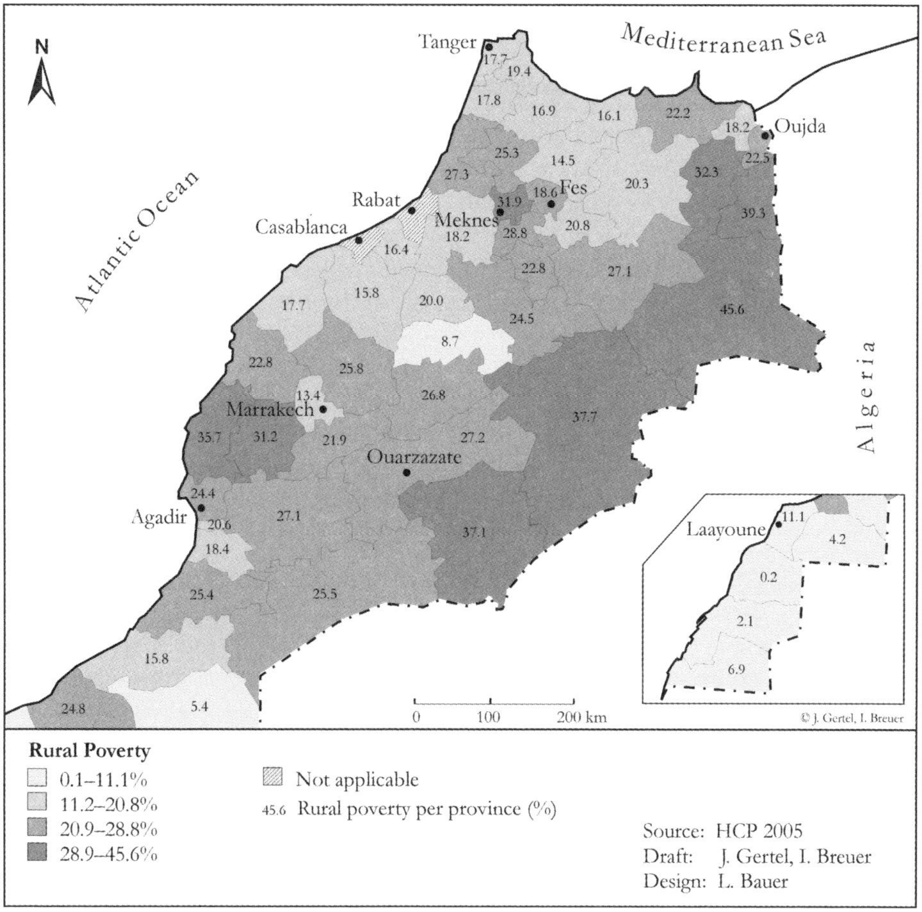

Map 1-4: Morocco – Rural Poverty

The eastern Moroccan high plateaus are, for example, dominated by the pastoral production of sheep and goats. Local pastoral livelihoods have, however, been subject to interventions by the state and international development agencies. These include the creation of cooperatives, the re-organization of pasture access, and the implementation of technical innovations (cf. Mahdi, this vol.). The region's pastoralism has thus undergone far-reaching social and spatial restructuring, such as sedentarization and the emergence of small towns (cf. Tag, this vol.), which lead to the emergence of new social groups of wealthy pastoral entrepreneurs and impoverished ex-nomads (cf. Rachik, this vol.). The region's economy is also shaped by informal cross-border trade in sheep and consumer goods and maintains a key role in supplying Morocco's urban centers with meat (cf. Khalil, this vol.). In Morocco's regional texture, the eastern plateaus remain one of the poorest parts of the country, although labor migration to the agricultural areas of southern Spain has recently become widespread.

The High Atlas – traditionally featuring multiple systems of ascending and descending transhumance – and the interior Pre-Sahara, differ in many respects from this. Only a minority of the population engages exclusively in pastoralism. Pastoral activities are articulated with other economic activities, such as trade, wage labor, and artisanship, many of which are performed in Morocco's urban areas. Most pastoral households are thus split up between the tent, the village or farm, and the city (cf. Breuer, this vol., High Atlas). A massive export of labor to urban areas, where men engage in temporal or seasonal labor markets, characterizes the local economy. The region also experienced early emigration to France in the 1960s/70s. Its repercussions are felt until today, for instance, when the French pension system guarantees social security for some early migrants. Recently, as international tourist agencies strongly market nomadic images, tourism also adds to the income of some pastoralists (cf. Werner, this vol.), while other places although connected to external markets, remain rather destitute (cf. Breuer, this vol., Pre-Sahara).

In the Middle Atlas and in the Oulmes mountains traditional pastoral systems were considerably altered by processes such as land privatization, social stratification among the herders, and intense lamb meat commercialization. Simultaneously, local territorial configurations in the agro-sylvo-pastoral system are increasingly shaped and threatened by the emergence of capital intensive forms of enterprises that dominate local policies and change the access to pastoral resources (cf. Aderghal, this vol.). Pressure on collective pastures is thus high, pasture degradation is serious, and a multiplicity of new herd mobility patterns have emerged (cf. Djoudi et al., this vol.).

The Souss, unlike other pastoral areas, is one of the economically most dynamic regions of Morocco, with comparatively low poverty rates. In the past decades it has witnessed massive growth of export-oriented agrarian production, a boom in the construction sector, and rapid urbanization. At the same time, it offers, with the endemic Argan forest, a unique space – partly protected as a UNESCO Biosphere Reserve – where sylvo-pastoral animal production is practiced. During the last years, the Souss has become the destination of many pastoralists from Morocco's arid areas, especially in times of drought (cf. Werner, this vol.). Their presence has lead to the emergence of extensive conflicts with complex constellations of interests and alliances (cf. Turner, this vol.).

Notes

(1) Pastoralism has been conceptualized in various ways, and multiple definitions have been developed. For instance, households are delineated as 'pastoral', if livestock or

livestock related activities account for a specified share, for example, at least 50% of the total gross income (which, in turn, comprises revenues from the value of marketed production and subsistence production). Households which obtain more than 25% but less than 50% of their gross income from livestock on communal grazing land and more than 50% from cropping activities are defined as 'agro-pastoral' (cf. Swift 1988). These conceptions go beyond the contested notions of nomadism and nomadic people (Gertel 2002). They offer the advantage of being applicable to the multiple contexts in which pastoral activities are carried out today, as, for example, to individuals and households who have been forced to destitution and depend considerably on non-livestock revenues, or to wealthy households who have successfully diversified into trade, transport, agriculture or government employment (cf. Meadows and Morton 2000; FAO 2001; Rass 2006).

(2) The number of poor pastoralists worldwide are estimated to range between 35 to 90 million, depending on the definition and measure of poverty applied (Rass 2006). For Morocco, the International Livestock Research Institute estimates the number of poor livestock keepers at 2,5 million people, a number comprising both rangeland-based and mixed (irrigated and rainfed) production systems (Thornton et al. 2002, 25). According to the same source, poor livestock keepers in rangeland-based systems amount to about 131.000 persons.

(3) The map is based on data from the 2004 "National Map of Poverty" (NMP), published by the Moroccan *Haut Commissariat du Plan* (cf. HCP 2005). We have reworked these data in order to calculate, for each province, poverty levels of the rural population only, while the urban population, generally characterized by lower poverty levels, has been excluded from the calculation. The map is based on administrative boundaries which, in many cases, are not congruent with natural zones distinguishing different pastoral phenomena. Poverty is defined as "the percentage of individuals with expenditure below the threshold of relative poverty, fixed at 1,687 Dh per month for an average rural household with 6.4 members." (HCP 2005, 12; our translation). This threshold represents an expenditure level, that in theory permits access to a minimum of calories (1,984 kcal per person and day) plus a certain amount of non-food items. However, expenditure data are available for a limited sample of the Moroccan households only. The communal poverty data displayed in the NMP are thus based on model calculations, the model itself refers to data from two sources: (1) The 2004 national census (RGPH), in which structural features of all of Morocco's households were recorded, and (2) to the 2000/01 household consumption and expenditure survey (ENCDM), in which detailed expenditure data are provided for a sample of Morocco's households. The poverty data were derived by taking into account regionally specific correlations between structural household features and household expenditure (cf. HCP 2005).

–2–

Mobility and Insecurity: The Significance of Resources

Jörg Gertel

Brahim – A Family of 'Truck-Nomads'

"Mhamd and I are nomads (irahhaln)," 33-year old Brahim relates.[1] *"All year long we live in a tent and take care of our family's herd of 250 sheep and goats. We are a family of nine – I have three sisters and three brothers. Two of my sisters are married and left our household to live with their husbands. My brothers Addi (29) and Idir (23) are migrant workers, going to places like Marrakech, Beni Mellal and Ouarzazate." Brahim explains: "My father lives in a large house in Oussikis with my mother, my unmarried sister and Addi's wife and son; they stay there because Addi can't take his family along when he looks for work in the cities. Idir and Mhamd are still unmarried and don't have to leave their families behind. I am lucky. My wife and my five-year old daughter live with me in the tent."*

"My father stays in Oussikis all year long and takes care of the fields, especially our 250 apple trees. Sometimes he also works on the fields of a large farmer in Oussikis. In summer we always come to pastures which are only about 15 km from the village. We like it, because then our contact with the family is very close. My father even comes and spends about two days per week with us in our tents. And he does the shopping for us on the weekly market. Often he also sells animals himself." Almost half of the family's income is derived from the sale of animals; the revenues from wage labor and sale of agricultural products constitute the rest. In years with a plentiful harvest the apple sale can amount to 20% of the total income.

After spending the summer near Oussikis, Brahim, his wife and daughter and his brother normally move to the distant pastures of the Saghro Mountains 90 km away. "These are the traditional winter pastures of our tribe, the Ait Atta," Brahim tells. "But in 2002 everything changed. In autumn there was hardly any rain at all. That is why the grass on the winter pastures couldn't grow. We didn't know what to do with our herd. Then we talked to other herd owners in the area. Some of them had decided to leave the region with their animals, trucking them to the southwestern Souss because down there, it had rained a whole lot. We heard that in the Souss the pastures were supposed to be great. In our family we discussed the issue and visited many other nomads who had been to the Souss or were planning to go, in order to find out more about the conditions. Then we decided to take a look ourselves. So my father, Mhamd and I drove down and looked around for two weeks. Finally, we found good pastures in the area around the Youssef Ben Tachfine lake, about 40 km from Tiznit. But from Oussikis this is really far, one and a half days by truck!"

Late in the autumn 2002, Brahim and Mhamd rented two trucks and transported their entire herd and also their tents to the selected area around the lake. "We stayed there until April 2003," Brahim remembers. "But for me, another change occurred. This time I also had to leave my wife and daughter behind with my family in Oussikis, because we thought the Souss would be very tough. But luckily we didn't have any problems with the locals, because our campsite was far away from any village and we moved it several times. Still, to buy groceries, we had to walk several hours to the market in Arbaa Ait Ahmed. In spring, when it got hotter, water became scarce. It was time for us to return to Oussikis."

Hmad – A Nomad and Remigrant

"Since my earliest childhood I have taken care of sheep and goats. Now I am going to tell you what I have done since. First, I was a shepherd. Since I was small I have tended camels, sheep and goats in the mountains and I was on many different pastures. We always drank sour milk and milk. Sometimes we also ate meat. [...] I went into the Saghro and the Bougafer with my father and mother and my children. At times we had a tent and also a house in Oussikis. We pitched our tent in the mountains. The camels and the herd also stayed there. [...] Now I live here in a village in Oussikis. I am 67 years old. [...] Living in a tent is good, but you also get tired and tent life is hard. The house is more comfortable and less harsh. There I have my peace."

In 1975 Hmad migrated to France, where he worked until 1989 in a mine close to Lille. During this period he earned, as he said, "more money than ever before or ever after." But the 14 years of hard labor have also left their lasting marks on him. He still is handicapped, struggling with respiratory problems: "The mine is ruled by hierarchy. The patron tells you what to do, and you just do it. You have no freedom. As a nomad I was my own master and did what I wanted. In the mine you earn a lot of money – as a nomad only little. In the mine the work is hard. It is dark and you can hardly breathe. I used to cough and was sick very often. [...] I was here every summer and I went to my herd in the mountains – for the peace and quiet. I slept in the tent and observed my animals. The air here is clear and we ate lots of meat. During my entire stay in France, I always wanted to return to the nomadic way of life."

In contrast to the majority of other emigrants of this generation, Hmad did not sell his herd before migrating. He employed a shepherd who took care of his herd. In addition, he constructed a solid shelter from stones on a nearby pasture, where he keeps his animals in winter. Compared to former times, his herd is much less mobile and stays the entire year within a 20-km radius of Oussikis: in winter the animals remain in the shed, in summer they are lead to two different pastures, which are both located a half-day march from Oussikis. Hmad rejects the idea of greater nomadic mobility for himself. In bad years he rather invests in supplementary feeds, mainly barley, which traders offer on the local markets. Hmad claims: "Nomads have changed. When I came back, the times had changed – even the people, the pastures, the nomads and agriculture. There is no more rain and no more graze on the pastures. Rather there is barley to feed the animals, and it simply has to be bought. [...]. And now the children want to go to school. In the past, the children took care of the herd. Time is moving on."

From the money he saved in France, Hmad bought 20 fields, making him one of the largest property owners in the village. He grows apples and potatoes, most of which he sells. He employs day-laborers from the village for agricultural work. Hence, his investments provide a comfortable extra-nomadic income. Working with the herd is not a pure hobby for him. Yet, he frankly acknowledges that he does not depend on his revenues from the herd, which only amount to a small fraction of the total household income. He himself loves tending the herd most, as he relates, although his presence is not required to complete the work. Up to now his 18-year old son Moha does the shepherding. But in the long term Moha does not want to remain a shepherd. Instead he wants to go to Casablanca in order to look for wage labor. Hmad understands that the flock must be sold when he himself is too old to herd and his son no longer shepherds.

Icho – An Impoverished Nomad

"My father was a nomad. He also had some fields, but my father leased the fields to other people, because we didn't know how to farm," Icho explains. "My father taught me the work with the herd, since he was a nomad. Once I was grown, I herded myself. I like the animals and the work, because one is always close to profits, this means money. [...]. I have never traveled. I've always taken care of the herd and the fields. I've sold animals and bought whatever we needed on the market. [...]. The most important thing for me is shepherding. I don't know how to do other work. I only know how to shepherd. But I have also spent the best moments of my life with my herd."

"When I was young, I always shepherded the flock. We were seven brothers. We had camels and some sheep. When I was 30 years old, we divided the herd amongst us brothers, and then I married. In winter we went into the Saghro, in summer into the Adrar and to Mjdi. In the past, you could sell animals and purchase fields with the money. My brothers and I, we bought eight fields. But these are divided now, and each of us only has very little land. When we divided the herd, I got 30 animals. I herded them till I finally had 437 animals. And then the drought (of 1982) came along, and after it I was back to 30 animals. Right after dividing the herd, I also had one female camel. In the course of years, I had 22 camels. The drought left me with only two."

"My father didn't let me go to France. I wanted to leave with Mora (a well known patron who was in the area to hire workers). My father said no. He wanted me to work with the herd instead of going to France. My brother was allowed to go. The rest of us had to stay. My father wanted it this way. My brother is still in France. He only visits us in his vacation."

"My two sons and two daughters never went to school. They only shepherd. They live in the 'khlâ' (literally emptiness, the temporarily inhabited pasture areas), in a cave. We go to the markets in Tinghir, Boumalne and Msemrir. In the past, I never went to the hospital. [...] My children want to continue being nomads, because they also don't know any other work. And they have no opportunity of going to France. You just have to try to survive and that's all. [...] My children like living in the mountains best, because there it is calm and quiet. Whenever I ask them to go to the house, they don't want to. They say: 'Even if we should have no more animals, we will stay here, because here we have water sources and firewood.'"

In 2004 Icho's situation changed dramatically. He fell very ill and had to be hospitalized for some time. The family did not possess the necessary financial means to pay for his treatment. They had to sell a third of the herd. One of Icho's sons gave up herding, and ever since seeks employment as a wage laborer "wherever he may find a job", and thus helps the family to meet its needs for cash.

Mobility as Strategy

The history of mankind is indispensably based on mobility, practiced by nomadic people right into late modernity. While nomadic ways of life perish, often under the conditions of insecurity and severe marginalization, complex and globally-integrated pastoral livelihood systems arise and new forms of mobility unfold. The three examples aptly illustrate the different mobilities that shape everyday lives of pastoralists in Morocco. Some still rely exclusively on animal husbandry for their livelihood, either through the use of traditional pastures, or more recently through the transport of animals to distant grasslands. Others employ additional sources of revenue, like agriculture or labor migration to Moroccan cities or abroad. Mobility is omnipresent: it is a skill to make use of pastures, and a prerequisite to becoming involved in distant labor markets. It can be described as a means to spread economic risks and – given a larger time frame – it entails hope to improve the lives of future generations. Hence, this paper explores mobility in a world that is increasingly characterized by the displacement of people, goods and information. Here, mobility is conceptualized as a strategy to maintain and enhance access to resources. While the term strategy comprehends a form of social practice, grounded in discourses of strategic decisions, the notion of resources is delineated as means that enable action.

Two bodies of literature shape the contemporary discourse on mobility. One takes a historical and developmental perspective on nomadic people, addressing particular situations in post-colonial and post-socialist countries (Beck 1988; Spittler 1998; Claudot-Hawad 2001; Finke 2005; Stammler 2005). Mobility, in this case often studied by social anthropologists and geographers, is regarded as a fundamental aspect of nomadic life, that helps to cope with insecurity in harsh environments (Scholz 1995; Schlee 2005; Leder and Streck 2005). The second corpus originates from a more western and sedentary perspective referring to societal changes in industrialized countries. Predominately investigated by sociologists, mobility is here viewed as a basic 'principle' of late modernity, transforming societies in the context of globalization (Urry 2000a, 2004; Bonß et al. 2004; Szersynski and Urry 2006; Beck 2004). It is argued that while both literatures offer valuable insights, they remain partial and disconnected in their analyses and fail to provide a comprehensive concept of mobility.

The relevance of mobility in the nomadism literature seems to be outstanding at first glance, as it allegedly symbolizes the ontological difference between nomadic and sedentary people. But the conceptual insights on mobility remain vague, as the term is predominantly used descriptively, and is restricted largely to spatial mobility. Moreover, the definition of mobility is often subordinated to the very understanding of nomadism.[2] Hence, this body of literature offers little theoretical insight into mobility, but rather provokes questions, about problems of inequality, such as the impact of spatial mobility in the context of livelihood security (Janzen 1999) and uncertainty (Scoones 1996). These questions are crucial in opening up an interface to conceptualize mobility, for example, as a (coping) strategy that depends on access to resources (cf. Breuer, this vol., High Atlas).

New sociological terminologies indicate a changing perception of society. Concepts like scapes (Appadurai 1991, 1997), socio-spheres (Albrow 1996), transnational social spaces (Pries 1998; Faist 2000), connectivity (Tomlinson 1999; Wellman 2000), fluids (Mol and Law 1994) and liquidity (Bauman 2000) capture recent transformations in the context of globalization. Urry explicitly argues for a 'mobile sociology' (2000a) and prepares the discussion for a 'sociology beyond societies' (2000b). Theorizing global processes, he is concerned with the mobilities of people, objects, and images, the interdependencies between them, and the social consequences of such mobilities. Indeed, he observes a 'mobility turn' in social sciences and argues for a 'mobilities paradigm' (2004). New theoretical insights center upon the notion of 'motility' (i.e. mobility potentials) and the cosmopolitan perspective.[3]

The cosmopolitan perspective is based on the assumption that mobility is a basic principle of modernity such as individuality, rationality and globality (Bonß 2004). Under the conditions of radicalized modernity the connection between social and geographical mobility breaks up. The readiness for spatial mobility is no longer a prerequisite or a guarantee for social mobility. New forms of social integration rather arise from telepresence and global connectivity. In the age of Internet individuals realize specific projects without physical contact, without being co-present or moving in space. Subsequently, Bonß et al. (2004) conceptualize mobility as an ambivalent phenomenon with two dimensions: movement and motility. "Whereas movement indicates the effective mobility performance, motility refers to the mobility potentials" (ibid., 13).

Kaufmann et al. (2004) further elaborate on the concept of motility as a link between spatial and social mobility. According to them spatial mobility refers to displacements of people, goods and information, i.e. the movement of entities from a starting point to a destination via a specific trajectory that can be described in terms of space and time.[4] During this journey the entities may experience a change in status (such as transforming identities or value adding[5]), while the points of departure, traversal or destination are simultaneously being shaped. Social mobility, in contrast, is captured as the transformation in the distribution of resources or social positions of individuals, families or groups in a given social structure or network (ibid., 747). Consequently, Kaufmann et al. emphasize the necessity of studying the potential of movement, with regard to possibilities, constraints and consequences of social and spatial mobility and argue for the invention of motility as the capacity of entities to be mobile. But instead of linking this capacity to the access to resources, they apply a circular argument stressing: "Generally, motility encompasses interdependent elements relating to access to different forms and degrees of mobility" (ibid., 750). Thus, mobility is presented as a resource by itself. Yet, it needs to be asked, what in the first place enables people to be mobile (i.e. the access to resources). Hence, this delineation of spatial and social mobility addresses crucial questions about the consequences of displacements and the relations between mobility and social positions, and thus exceeds the largely descriptive concepts applied in the literature about nomadic people. Yet a systematic investigation into the analytical dimension of strategies and into the capability of action is missing.

While the literature on nomadism 'undertheorizes' the notion of mobility, it offers a problem-oriented perspective on unequal development and a contextualized view on insecurity, comprehending spatial mobility as a coping strategy. The literature on late modernity in contrast offers theoretical insights into a globalizing mobile society, but 'undercontextualizes' the empirical dimensions of mobility. It largely restricts itself to situations of affluence in the northern hemisphere, and thus fails to understand the relation between social reproduction and mobility under the conditions of insecurity.

This paper focuses in particular on the situations of insecurity, prevalent in less industrialized countries like Morocco. It assumes not only a fragmentation of development, represented by a growing rift between the wealthy and poor, but also a post-colonial development within which the mobility of pastoralists is shaped by the consequences of late modernity, such as time-space distanciation, disembedding, and reflexivity (Giddens 1990). Subsequently this paper defines mobility as a *strategy* to maintain and enhance access to *resources*. It argues that a thorough investigation of mobility has to include three elements:

First, a focus on actors and their means of action within social structures. Here, households as the smallest units of social reproduction are the object of analysis. Correspondingly the resource structure of households shaping the possibilities of mobility (e.g. movement of herds, displacement of labor) is discussed. This discussion draws on development studies and elaborates on notions of vulnerability, livelihood security and concepts of capital and resources.

Second, a focus on the territorial scope of action. This dimension of mobility is conceptualized beyond a simplified notion of spatial containers with fixed boundaries (e.g. nation states). It draws on debates in geography and political sciences about the constructed nature of regions (e.g. notions of scapes and locality).

Third, a focus on the temporal scope and the social consequences of mobility. Here, identity in its negotiated aspects is addressed (e.g. 'Berber' or 'Ait Atta' in France). The discussion refers to cultural studies about discursive constructions of meaning and their materialization (e.g. notions of appropriation and embodiment).

In the following an analytical frame, which accommodates an empirical analysis of mobility, based on the aforementioned elements, will be developed. The term strategy will be specified and the notion of discourse explored. Since the term strategy is derived from military and business environments and connotes active and planned agency, it is contested whether its application is justified in household studies. Yet strategies and strategic decision-making are part of social practice and require detailed analysis (Wallace 2002). Against this background Hendry (2000) comprehends strategic decisions as part of a discourse, as a body of language-based communications that mediates between cognition and action. Operating at both the structural level of social reproduction and the agency level of intentional communication, the strategic discourse constitutes "the medium through which choices are discussed and recorded, interpretations developed and expressed, and strategic actions initiated, authorized and acknowledged" (ibid., 955). Hence, mobility is thus not just there. Complex processes, linking cognition and action, structure the forms of mobility. These processes depend on access to resources (cf. Peluso and Ribot 2003), on the context of social reproduction, and on the discourses at work.

When shaping strategies, discourses largely exceed the household level. Discourses are heterogeneous linguistic events, which are not penned by one single author. Yet, they do not represent free-flowing constructions, but are always embedded in society, are often institutionally reproduced, and produce material consequences (Barker et al. 2001). The relevance of discourses rests on the fact that they structure access to resources. For example, the question: 'is nomadic mobility necessary', requires an inquiry into the means of representation (Hall 2002). Subsequently one has to ask: who is opening up this topic, who is questioning whom, and who is responding to whom? Do all actors – for example, nomadic people, bureaucrats or scientists – have the same 'right to narrate' (Spivak 1988). Or differently asked: who has the possibility and the authority to define terms and to occupy themes; men or women, literates or illiterates? Furthermore, it has to be examined how discourses materialize, for instance through the codification in plans and in sedentarization programs for nomads. A debate about pastoral mobility may thus thoroughly change the conditions of pastoralists, without anyone ever having heard their voices.

Hence, mobility – as a strategy to maintain and enhance the access to resources – is discursively constructed and embodied in the context of social reproduction. It materializes during action, shapes territorial structures, and may in the long run foster (un-)intended consequences, such as social mobility and new identities. The truck nomad Brahim is an example of this. Everything changed for him in 2002, when the drought set in and he and his family negotiated and decided to leave their traditional pasture in order to transport the animals hundreds of kilometers away. They tremendously changed their scope of spatial mobility, entering a new social context of interaction in a foreign environment. Simultaneously they altered their household composition, at least temporarily, to secure their social reproduction.

Mobility in Situations of Insecurity

Mobility unfolds in an unequal world. Despite tremendous efforts in the past 50 years inscribed into the agenda of development, economic disparities are ever increasing between and within countries (cf. Milanovic 2002). The contradictions of the global economy are placing Morocco, for example, at a critical juncture for desperate African migrants on their way to Europe, while processes of economic and social polarization are shaping pastoral livelihoods inside Morocco (Davis, this vol.). In situations of insecurity, spatial mobility is one of the key strategies to access resources, such as good pasture or rewarding employment (cf. Chiche; Werner; Breuer, all this vol.).

Starting in the 1980s the literature on development was marked by a shift towards theories addressing insecurity by employing a bottom-up perspective, focusing on subsistence production, reproduction crises and resource-portfolios and thus aiming at a more detailed analysis of social practices. Within this debate three concepts contributed to a deeper understanding of mobility as a (coping) strategy; namely approaches about survival economies, vulnerability, and livelihood security. These concepts share interest in analyzing interactions between the local and global, and emanate from empirical household studies.

The concept of survival economies assumes that the 'group of the insecure' (Schicht der Ungesicherten), such as impoverished pastoralists, apply various strategies in order to secure their livelihoods (Evers 1987; Schmidt-Wulffen 1987). Since this group's prime intention is security, its members try to spread (economic) risks rather than to maximize profits. For example, households combine various sectors of production to reduce insecurity, i.e. income and goods might be obtained from wage work in informal and formal sectors as well as from subsistence production. In addition, new mechanisms of solidarity are developed and existing personal social networks are reinforced. Spatial mobility, such as labor migration, is also part of the strategies to survive under conditions of insecurity. These activities are not necessarily attributed exclusively to entire households, but also to individuals.

> The individual becomes a free-floating agent, acting daily, in different spheres, with different keys of action. To achieve prosperity people have to be successful on all occasions. An individual who tries to integrate all such different ways of being into one frame of action has not understood that modernity demands a lot, not least a multiplicity of human actions (Semsek 1987, 283).

In contrast to conventional views of national economies, sectors or economic cycles, the bottom-up-focus on households or on 'units of reproduction' (cf. Smith et al. 1984) emphasizes small analytical categories. This is guided by the assumptions that even individuals can produce with both a subsistence and a market orientation, that they are integrated into different social networks and are able to act in spatially mobile ways. Hence spatial mobility is explicitly regarded as a risk minimizing strategy.

The case studies of Hmad, Brahim and Icho reveal these complexities. A pastoralist can be involved in several economic activities at once, and revenues from different family members are pooled in a joint household. Here, risks of social reproduction are redistributed among generations. The young and the old, who either cannot work yet or not work anymore, as well as women and children who often perform unpaid herding or household tasks, render indispensable assistance to the household economy, albeit they do not directly participate in earning of monetary revenue. However, especially the elderly may dispose of significant transfer incomes, since they may have access to state welfare services or even receive pensions from foreign countries.

This analytical perspective significantly structures development research. Yet there are two critical points. The term action is neither delineated nor explicitly integrated into a theory of action. The assumption of rational action based on a given purpose (Zweckrationalität), connotes continuous cost-benefit calculation, which appears to be too reductionist. Already Weber ([1921] 1972, 13) comprehends 'acting' to be more than simply guided by an utilitarian rationale. He points out that action, from a heuristic viewpoint, can be considered as motivated not only by purpose-based rationality, but also by a rationality based on values and affect. Thus, a single motivation for action hardly exists. Moreover, the importance of resources remains unexplained, as does the question of how strategies are generated. A relational perspective into the scope of action, determined, for example, by volume and structure of an asset portfolio is still missing. Before further exploring these points of critique, the perspective of the survival economy will be extended by discussing the vulnerability and sustainable livelihoods approaches.

The concept of vulnerability is rooted in Chambers' critique of the notion of poverty in the development discourse (1989). He emphasizes that the concept of poverty has become a catchall term, concealing social differences and perpetuating stereotypes. He stresses that it is based on the assumption of an amorphous and undifferentiated mass of poor individuals. Chambers differentiates between several dimensions of deprivation. Aside from poverty, he lists isolation, powerlessness, and vulnerability. In so doing, Chambers insists that vulnerability is not the same as poverty. It means not lack or want, but defenselessness, insecurity, and exposure to risk, shocks and stress (1989, 1).

> Vulnerability here refers to exposure to contingencies and stress, and the difficulty in coping with them. Vulnerability has thus two sides: an external side of risks, shocks, and stress to which an individual or household is subject: and an internal side which is defenselessness, meaning a lack of means to cope without damaging loss (ibid, 1).

Hence, the group of the poorest is not necessarily identical with the most vulnerable. Drawing on these considerations, Watts and Bohle (1993) identify three risks for individuals and groups, which they portray as the fundamental aspects of vulnerability: the risk of exposure to crises, stress and shocks; the risk of inadequate coping-strategies; and the risk of severe consequences and impaired capabilities of recovery. Vulnerability is understood as a multi-layered and multidimensional realm of activity, which is defined by economic and institutional choices and constraints of individuals and groups.[6]

The concept of vulnerability also allows to stress temporal dimensions of reproduction problems, for example, in the context of droughts. Accordingly different phases are distinguished, such as baseline vulnerability, current vulnerability, reproduction crisis, and recovery (cf. Bohle et al. 1991; Gertel 1997). For pastoralists a drought may have different intensities of distress: it may contribute to an underlying trend within baseline vulnerability; or it may work as a trigger to translate baseline into current vulnerability; or ultimately it may become effective as a shock fostering severe consequences of reproduction, such as food crises. Although long-term developments (e.g. baseline vulnerability) often combine with short-term conjunctural events (e.g. current vulnerability), analytically they can be differentiated. In sum, this concept allows for the analysis of the fine-tuned coping strategies applied by nomadic people. In situations of increasing insecurity even impoverished pastoralists try to avoid depleting all their resources at once, but rather – by envisaging different time frames of action – they seek to design specific strategies (Wiese 2004). Here the livelihood concept is useful, as it investigates scopes of action according to the asset portfolio at hand (cf. van Dillen 2004).

The livelihood-concept was coined in the early 1990s and like the vulnerability approach largely originates in the work of the Chambers-School at the Institute of Development Studies (IDS) (cf. Chambers and Conway 1991). The widespread use of the concept was fostered by the Department for International Development (DFID).[7] Livelihood is comprehended as follows:

> A livelihood comprises the capabilities, assets (including both material and social resources) and activities required for a means of living (DFID 2000, 1.1.).

Disposable assets are at the center of the analysis by means of which actors are capable to achieve their livelihood (Carney 1998). Assets – according to the model's considerations – emanate from different kinds of capital, namely natural, financial, social and human capital (DFID 2000; Scoones 1998). The analysis of the capital endowment is embedded into the analysis of the respective vulnerability context, and focuses also on institutional processes and organizational structures. From this the livelihood strategies result (e.g. migration, or changing household composition). They determine how sustainable a livelihood finally is (cf. Scoones 1998; Ellis 2000).

Three points of critique apply: First, the livelihood approach enables the identification of coping strategies in situations of insecurity. But it does not provide insights into the causes of risk generation and reasons of economic inequality. Societal structures and macro-economic conditions are treated as residual categories. Hence, the causation of (development) problems, such as the deprivation of Icho and his family, is implicitly ascribed to the individual. Literature about survival economies (Evers 1987; Smith et al. 1984), in contrast, presents a more comprehensive analysis of poverty, addressing historical and structural causes of unequal development. Second, the livelihood approach is characterized by a certain distance to theory. Its theoretical weaknesses are exemplified by the absence of concepts addressing human action, like Giddens' understanding of resources as a key variable in his theory of structuration ([1984] 1995) or like Bourdieu's definition of capital in his theory of practice (1972). Key categories of the livelihood approach like assets, resources and capital are not clearly delineated, but interfere with each other and as such they dissolve. The difference between observable action and structural conditions of the action are not revealed and the two analytical levels are not differentiated. Third, in spite of a contrasting intention, the livelihood concept is limited in its analytical scope to economic possibilities. At the center of the concept are assets, and especially aspects of capital endowment are researched. The understanding of livelihoods is thus reduced and shortened conceptionally and empirically to the monetary level, probably for reasons of measurement because corresponding 'factors' are more easily counted and quantified. Considerations from rational choice approaches (e.g. Coleman 1990; Johnson 1997) inspired these arguments. Therefore, it is important to expand the conceptual understanding of human action and to determine how mobility strategies are enabled and constrained by households, their resource structure, and social reproduction.

Mobility and Social Reproduction

Guided by Hendry's (2000) analysis of strategic action and by Giddens' (1995) theory of structuration, social reproduction is comprehended as a recursive course of events related to the restoration of structural properties in social systems.[8] Social systems are conceptualized as "the patterning of social relations across time-space, understood as reproduced practices" (Giddens 1995, 377). Therefore, the starting point for social reproduction is social relationships, which in terms of institutionalized practices are always recursively reproduced anew, often along a routinized performance.

In this framework, households are a center for social and economic reproduction, constituting reproduction units (Wong 1984; Gonzalez de la Rocha 2006). In addition to an analytical unit, they are a social unit in practice (Wallace 2002). The term 'unit' does not refer to an internal homogeneous or to a self-contained structure; it only indicates commonalities, which the members of a household share. Friedman points out that the work of one household member cannot be assessed in isolation from the work of other members: "In fact, the value of each form of total household income is a product of its being in relation to the other forms of income. The total household consumption fund is composed of a variety of labor relations that are drawn together, put into systematic relationship with each other, and derive significance only from being linked to one another. Householding practices are those which link these in a totality" (1984, 46). Correspondingly, the internal structure of households is constituted by different patterns, varying in economic, social and demographic aspects (cf. Breuer, this vol., High Atlas). Moreover, individual members do not always have the same interests. Differences, for example, due to gender and age may enhance conflicts about lifestyles or the use of resources. An overview of all income and expenditures is often not possible even for the head of household (Rogers et al. 1990). The social forces and relations of reproduction within households are captured by Harriss et al. (1990, 284):

> Forces of reproduction comprise the quantity and quality of labour (affected by household size, composition and health status), the means of consumption (e.g. food) and the means of reproductive work (water, housing, technologies for food preparation, for household maintenance and for biological reproduction). Relations of reproduction comprise the gender and generational division of domestic labour, decision-making power and control over resources and over their transfer between generations and households. In this framework, nutrition and health status is simultaneously an outcome and an input into the process of production and reproduction, as are stocks of goods and of money.

In this context of social reproduction, strategies – such as mobility – are developed and discussed by actors, and are authorized and conducted. Corresponding action evolves in reference to the possibilities of a household. Strategies remain, of course, ambivalent, performed under conditions of power structures inscribed into the specific social context. Mobility as a strategy, with its temporal and spatial manifestations, is thus not an outcome of linear decision-making-processes, and does not result from a single cause. On the contrary, it is grounded in cognition, translated into discursively contested actions, that again reflect earlier social negotiations. Hence, strategies represent nodes of ongoing decision-making processes, largely determined by the resource portfolios of reproduction units (i.e. households).

Mobility and Resources

The notion of resources – transformed from the ancient idea about reciprocal and regenerative relationships between humans and nature (Shiva 1999) into an utilitarian concept of 'inputs', as is the case in the conventional livelihood approach – refers here, more generally, to the capability of doing things. Social action emanates from needs, which, stored in different levels of consciousness, generate a certain potential of action (Giddens 1995). Social action can be explained by actors themselves – at least to a certain extent – and it is under their reflexive control. Actors continually monitor the flow of their own activities and also expect this from others; they routinely also control the social context in which they move. However, the consciousness of human actors is limited: by the unconscious, by unrecognized conditions of action, and by unintended consequences, which precede or succeed the reflexive control of action. Giddens claims that human history is created by intentional activities, but it is not an intended project in itself. He stresses the precondition for action and the crucial role of resources:

> Even the view that for an event to count as an instance of agency, it must be intentional only under some description or another is wrong. It confuses the designation of agency with the giving of act-descriptions; and it mistakes the continued monitoring of an action which individuals carry out with the defining properties of that action as such.

> Agency refers not to the intentions people have in doing things but to their *capability* of doing those thing in the first place [...] (Giddens 1995, 9; emphasis added).

The stress on 'capability' refers to the importance of resources, because it is resources – and more precisely access to and the control of these – which create the prerequisite for action. Subsequently, the structures of social systems are of central importance, because they are composed of 'rule-resource complexes':[9]

> The social systems in which structure is recursively implicated, [...], comprise the situated activities of human agents, reproduced across time and space. Analyzing the structuration of social systems means studying the modes in which such systems, grounded in the knowledgeable activities of situated actors who draw upon rules and resources in the diversity of action contexts, are produced and reproduced in interaction (Giddens 1995, 25).

Giddens differentiates between authoritative and allocative resources: the first emanate from the coordination of actions of human beings and the latter from control over material products or certain aspects of the material world. Allocative resources include three elements: material features of the environment (i.e. raw materials); means of material production and reproduction (i.e. technology); and produced goods, created by the interaction of the first two elements. Authoritative resources are also composed of three elements – the organization of social time-space, the production and reproduction of the body, and the organization of life changes. Giddens argues that any co-ordination of social systems across time and space necessarily involves a definite combination of these two types of resources (1995, 258). He further stresses that material resources, although fundamental to the expansion of power, cannot be developed without the transmutation of authoritative resources. Hence, analytically the interrelation between cognition, action and structure in respect to resources becomes apparent and is well suited to conceptualize mobility as a strategy. However, Giddens' concept has weaknesses. He is not addressing the meaning of resources within (economic) exchange systems. His notion of allocative resources perpetuates rather implicitly assumptions of neoclassical theory, overemphasizing material exchange systems, while neglecting symbolic aspects (e.g. gifts, reciprocity and the like). Although he distinguishes material and immaterial aspects, the discursive construction of everyday life – and thus of strategies – remains underestimated. Moreover, he does not comprehend the body as part of material resources. Subsequently the bodily consequences of uneven development in the form of poverty or malnutrition escape his analysis. Bourdieu's work, shaped by his experience in Algeria, becomes useful here, and in particular his notion of capital.

Bourdieu places his argument alongside a critique of conventional economic theory.[10] Economic theory, according to Bourdieu, depends on a definition of capital that represents and reproduces the historical invention of capitalism. This concept of capital – a master narrative – reduces the universe of exchanges to commodity exchanges, which are objectively and subjectively oriented toward the maximization of profit, and structured by economic self-interest. Subsequently economic theory defines all other forms of exchange as non-economic, and therefore altruistic. Bourdieu instead argues for a reintroduction of an expanded notion of capital. He comprehends capital as accumulated labor, whether in its materialized form or its incorporated, embodied form. He stresses that the accumulation of labor requires time, internalized or objectivated time. Capital exists in different forms: as economic, social and cultural capital.[11] He explains that economic capital is directly convertible into money and is explicitly suited to be institutionalized in the form of property rights. Cultural capital exists in three subtypes: as embodied, objectified and institutionalized cultural capital.[12] Social capital, finally, is made up of social obligations and is equated with resources based on group membership, on networks of influence and support (1983).

Instead of analyzing societies in terms of classes, Bourdieu uses the concept of field. A field is a system of social positions structured in terms of power relations. Relative power, expressed as position within this social grid, is determined by the quantity and structure of available capital (economic, cultural, social, and symbolic capital). Hence, the interrelations between households and society at large can be investigated in empirical analysis addressing the distribution and structure of capital (i.e. resources); and it can also be examined how a concrete resource portfolio – shaping the capability of mobility – delineates social positions within society. Yet, a social field translates into very concrete meaning.[13] Bourdieu stresses the embeddedness of the economy in society (even priceless things have their price), traces its articulations down to the body, and highlights possibilities of converting different sorts of capital. However, the definitions of his core categories are ambiguous (e.g. social and cultural capital) and his concept of capital is based on a theory of value that remains tied to the notion of labor as the universal explanatory variable. This results in an overgeneralization of economic processes (cf. Pribram 1998).

Juxtaposing Giddens' insights into rule-resource complexes and Bourdieu's concept of capital, in the following the term resources will be re-defined and applied to analyze social reproduction. The notion of resources comprises links to biology (e.g. bodies ultimately die), to labor[14] and property rights (e.g. accumulated income from the past can be transferred into present livelihoods), and thus to the discursive construction of rules (e.g. encoded in the discourse about private ownership vis-à-vis the commons). Based on these considerations, four forms of resources are distinguished (Gertel 1997; 2005b): (1) Incorporated resources (e.g. nutritional and health status), which are bound to the body; (2) Socially institutionalized resources (e.g. social networks), which are related to the subject; (3) allocative resources (e.g. land, machinery), which are linked to property rights; (4) monetary resources (e.g. savings), which also depend on property rights, but which can be more easily exchanged between people. This concept of resources offers the possibility of linking the notion of uneven development to the status of physical bodies. It also provides insights into social causes of reproduction problems and thus opens a new perspective to understand mobility as a strategy.

Incorporated resources are resources, that are directly connected with the physical body of the person. This refers primarily to physical and mental health and the ability of a person to work, but also includes educational aspects. Incorporated resources are not transferable to another person,[15] and are lost upon death. For the individual the preservation of physiological and mental capabilities is of fundamental importance.

Institutional resources are resources, that are outcomes of social relationships. Social relationships, expressed in institutionalized networks, result in context related claims,[16] such as claims for information, services and money. Networks can be personal social networks, used, for example, to exchange information or to insure that members receive help in case of an individual emergency. Networks can also be formally institutionalized forms of social contracts. Institutional resources are connected with a specific person, not with the physical body, but in the sense of a formal or informal contractual relation. Benefits cannot usually be claimed by someone not known by the contract partner; however, third parties known to the institution may under certain conditions claim them.

Allocative resources are raw materials, means of production and goods. Allocative and monetary resources are not anchored within the body and are not bound to the transitoriness of social relationships, only in the sense that they are based on property rights.

Monetary resources, i.e. monies, are, therefore, the resources that can be most easily exchanged between people.

This perspective offers several advantages. For one, the model of the Giddensian definition of structure, as a recursively organized complex of rules and resources, reveals the connection between individual actions and social structures on the conceptional level. By means of this tool, those societal processes can be registered which generate the conditions of action for the individual subject. On the other hand Bourdieu's concept of capital permits a comprehensive view of exchange processes. The resource concept eventually makes complex economical and social processes accessible to empirical analysis. Mechanisms, such as accumulation, mutual convertibility, the utilization of different resources by third parties, become qualitatively and quantitatively measurable. In addition the consequences of actions shaping the structural conditions of livelihood security can be linked back to the bodies of those affected. Correspondingly, strategies may also be reflected in somatic sensitivities and incorporated resources of the individual. Here bodies are conceptualized as the final instance of societal space. However, the discursive construction of experience – inscribing spatial and social mobility – still remains unexplained.

De-Localizing Mobility

In the context of globalization, 'having experiences' becomes increasingly less bound to situations of co-presence. Mass migration and mass media foster new diasporas, like 'Turks' in Germany or 'Moroccans' in France, and create so-called 'transnational social spaces' (Faist 2000; Pries 2001). They also induce the global dissemination of standardized pictures and images, which alter the conditions of experience. Scott (1991) notes that subjects are integrated into multi-faceted, often contradictory discursive systems, and are at least partially constituted by these. Experiences – as a 'nomad' or with 'nomads' – are thus not generated in neutral fields, but are marked by specific traces of domination. The concrete subject position has to be understood as result of discursive negotiation.

> Make visible the assignment of subject-positions (...) [is] trying to understand the operations of the complex and changing discursive processes by which identities are ascribed, resisted, or embraced, and which processes themselves are unremarked and indeed achieve their effect because they are not noticed (Scott 1991, 792).

Still, actors are not simple victims of discourses. Although embedded in concrete conditions (ibid., 793), they possess agency, which opens possibilities for choices even if these are limited. This aspect marks a crucial interface between Giddens' and Scott's argument, because both tie the ability to act to resource access. The acting subject in its physical embodiment is the very location in which the various possibilities to access material and immaterial resources in the context of discursive negotiation interlock time and again. Starting from there, the space for the interpretation of experience articulates itself.

In the context of cultural globalization Appadurai (1997) observes a new social practice: imagination. He argues that imagination escapes its special expressive realm of art, myths and rites, and became part of daily mental exercises of ordinary people. In contrast to fantasy, which rather emphasizes the private and individual, imagination creates ideas, connotes project-orientation, and consequently is a basis for collective action. He concludes that imagination has become the property of collectives, which he calls 'community of sentiment' (ibid., 8). Groups of individuals, who never met in face-to-face contact, start to think of themselves as, for example, Indians or Malays. Linked to cultural differentiation processes, Appadurai is thus particularly interested in the process of de-territorialization of interactions. Furthermore Appadurai contends that within the current global economy, fundamental disjunctures occur between economy, culture and politics. For their analysis he develops a concept that contrasts the territorial rootedness of interactions: "I propose that an elementary framework for exploring such disjunctures is to look at the relationship among five dimensions of global cultural flows that can be termed (a) ethnoscapes, (b) mediascapes, (c) technoscapes, (d) financescapes, and (e) ideoscapes. The suffix -scape allows us to point to the fluid, irregular shapes of these landscapes, shapes which characterize international capital as deeply as they do international clothing styles. These terms with the common suffix -scape also indicate that these are not objectively given relations that look the same from every angle of vision but, rather, that they are deeply perspectival constructs, inflected by the historical, linguistic, and political situatedness of different sorts of actors: nation-states, multinationals, diasporic communities, as well as subnational groupings and movements (...), and even intimate face-to-face groups (...). Indeed, the individual actor is the last locus of this perspectival set of landscapes. For these landscapes are eventually navigated by agents who both experience and constitute larger formations, in part from their own sense of what these landscapes offer" (ibid., 33).

The term landscape becomes a metaphor; it no longer denominates a circumscribable territory with solid boundaries, but rather dimensions of flows. These emerge from intrinsically perspectival constructions of diverse agents, each of which with a unique history. The decentralized individual subject is identified as basic instance, and the imagination as central social practice in the design of de-territorialized social spaces:

> These landscapes thus are the building blocks of what (...) I would like to call imagined worlds, that is, the multiple worlds that are constituted by the historically situated imaginations of persons and groups spread around the globe (...). An important fact of the world we live in today is that many persons on the globe live in such imagined 'worlds' (and not just in imagined communities) and thus are able to contest and sometimes even subvert the imagined worlds of the official mind and of the entrepreneurial mentality that surround them (ibid., 32).

For Appadurai spaces of interaction are not bound to a contiguous territory, thus 'the local' can be rethought. The term no longer connotes simply periphery in a societal power structure, and is no longer a territorial unit bound to a 'there'. The local rather designates everyday practice – at the centers of power as well as simultaneously at different locations. Emerging from imagination, networks of social relations define everyday practices. Subsequently (nomadic) mobility is not only to be conceptionalized as spatial mobility of persons moving between places, or as social mobility of people changing social positions. Instead, mobility has also to be examined in the sense of identification within new social landscapes, ultimately expressed in new identities. Nomads such as Brahim, Hmad, and Icho become truck-nomads, remigrants or utterly impoverished pastoralists, conducting different modes of spatial mobility, in different spheres, changing social positions and ultimately negotiate new identities.

Conclusion

Mobility in late modernity presents itself more than ever as a complex phenomenon requiring thorough investigations based on comprehensive analytical concepts and intense empirical research. This paper delineates mobility as a strategy to maintain and enhance access to resources; as a strategy that is discursively constructed and embodied in the context of social reproduction, that materializes during action, shapes territorial structures, and bears consequences, such as new identities. Households are the locus where the ability of spatial displacements intersects with labor, health and nutritional status of its different members. Conversely, the individual subject is to be considered as the last spatial and temporal instance of globalizing social struggles. While social interactions, experiences and identities are increasingly negotiated in social spaces, that unfold in fragmented and spatially disconnected territories, nomadic peoples and pastoralists cannot evade these developments. Hence, the discursive merging of a mobility strategy, its subsequent materialization, and its ultimate locking up in new social positions, has to be studied at the empirical level for individual cases. This volume aims to bring these experiences together.

Notes

(1) I am very grateful to Ingo Breuer for using the case studies originating from his research in Oussikis (cf. Breuer, this vol., High Atlas). I also would like to thank Enne Haehnle and Petra Kuppinger for their insightful and constructive comments on this paper.

(2) A wide range of definitions is at work: One position assumes that nomadism – as a principally different mode of society (Scholz 1995, 20) – is indivisible. Subsequently, a person can only be either a nomad or a 'non-nomad'. The opposite position is rooted in the assumption of the divisibility of what is represented by the term "nomadic". This notion no longer relates to the ontological status of a person, but rather defines processes that are rooted in everyday practices (cf. Gertel 2002). Accordingly a person can perform both nomadic and other forms of work and life (Tietze 1998). While the first definition assumes that nomadic people are the very embodiment of spatial mobility the second one allows to attribute other forms of the spatial displacements, such as labor migration, which is also performed by nomadic people.

(3) http://www.sfb536.mwn.de/veranstaltungen/B3_Workshop_0104_Dokumentation.pdf

(4) Forms of spatial mobility of people range, for example, from nomadism, residential mobility, migration, and travel to routinized everyday displacements.

(5) The discussion about global commodity chains offers insights into the displacement of goods and its social consequences (cf. Hughes and Reimer 2004; Fold and Pritchard 2005).

(6) In order to determine risks defining vulnerability, different concepts with different scopes can be consulted (cf. Bohle and Watts 2003; Krings and Müller 2001; Bankoff et al. 2004).

(7) Only a few years after the DFID-initiative hundreds of studies, based on the livelyhood concept, had been published (compare www.livelihoods.org): Its employment reaches from nomadism research to the analysis of urban poverty and pertinently marks the development discourse (cf. Scoones 1995; Meadows and Morton 2000).

(8) Giddens states: "In structuration theory 'structure' is regarded as rules and resources recursively implicated in social reproduction; institutionalized features of social systems have structural properties in the sense that relationships are stabilized across time and space (1995, xxxi)".

(9) Giddens emphasizes that rules cannot be conceptualized apart from resources. "Rules have two aspects (...). Rules relate on the one hand to the constitution of meaning and on the other to the sanctioning of modes of social conduct" (1995, 18). He considers rules "as techniques or generalizable procedures applied in the enactment/reproduction of social practices" (1995, 21).

(10) The following discussion is predominantly based on Bourdieu (1983).

(11) Later he adds other forms of capital, such as symbolic and linguistic capital.

(12) Bourdieu (1983) distinguishes: (1) The embodied subtype. This is where cultural capital is embodied in the individual, usually through socialization. Embodied cultural capital is not transmittable instantaneously like a gift, but rather strongly linked to one's habitus. (2) The objectified subtype. Things which are owned, such as scientific instruments or works of art. These cultural goods can be transmitted physically (sold) as an exercise of economic capital, and symbolically as cultural capital. (3) The institutionalized subtype. This is institutional recognition of the cultural capital held by an individual, most often understood as academic credentials or qualifications. This is mainly understood in relation to the labor market. It allows easier conversion to economic capital by guaranteeing a certain monetary value for a certain institutional level of achievement.

(13) He emphasizes, for example in 'Distinction' (1984), that the structure of the distribution of the different types of capital at a given moment in time represents the immanent structure of the social world.

(14) Not all resources are dependent on labor, as Bourdieu suggested in his concept of capital. Incorporated resources, for example, are expressed in the nutritional and health status of a person, that are not shaped by labor.

(15) While education can be socially accumulated in form of diplomas from educational institutes and titles (Bourdieu 1983), health and physical labor power cannot be accumulated beyond a certain limit.

(16) Swift (1989) introduces the term claims into the vulnerability debate, establishing relationships between the expected reciprocity within social networks, unofficial redistributive taxes and state redistributive transfer benefits.

–3–

History of Mobility and Livestock Production in Morocco

Jeanne Chiche

Regional analysis of the mobility of people and their herds, their livelihood strategies and living standards remain crucial for an understanding of pastoralism in Morocco. Over the course of the twentieth century regional processes have effected changes in habitat, in economic activities, especially in land cultivation and techniques of livestock production, and in the spatial mobility of persons, including mutations in the geographic direction and in the aims behind these movements. This paper addresses all of these matters and argues that mobility and livestock production in Morocco are deeply influenced by diverse livelihood strategies. The latter range from nomadic livestock production to immobile forms of herding associated with more permanent settlement, agricultural modernization and labor migration.

Disappearing Tents and Changed Mobility

Generally speaking, pastoral livestock production seems to have declined significantly in the course of the 20th century (cf. Maps 3-1, 3-2).[1] This is true whether we count the number of nomads or estimate the size of their herds and flocks. In 1931, about 16% of all Moroccan households lived in tents; today, tents are only used by transhumant pastoralists accompanying the herds, and sometimes as temporary habitat by workers on large construction sites. In the past, tents as a type of habitat were utilized for different reasons, such as pastoralism, or labor migration towards agricultural areas or towns (cf. Map 3-3). Pastoral livestock production and the use of tents dominated in the Middle Atlas, and in the sub-Saharan and Saharan steppes. Other tent dwellers – less numerous – can be identified as members of *guichs,* which at the time were not settled in permanent habitations.[2]

In the Atlantic plains, the main area of colonial agricultural production, the major part of the tents sheltered families from the central north of Atlantic Morocco, whose mobility had been enforced by the demand for workers. Among the Chaouia, southeast of Casablanca, for example, 35,956 tents were counted in 1931, whilst their territory had already been completely transformed into cultivated land. These migrants camped at the periphery of large towns or near large farms. But in many cases, the tent was also the refuge of former shepherds or *khammes,*[3] who were forced to seek employment with farmers in order to survive in the context of sedentarization and the expansion of cultivated land, which had been promoted since the beginning of the French Protectorate. These people, neither owning land nor large herds, and lacking other means of capital, worked as farmhands in exchange for the right to set up a temporary hut for the duration of their contract. These types of habitat were particularly visible in the Atlantic plains and plateaus. It was only in the course of the 1930s that they were progressively abandoned. Remarkable remnants of these habitations still persist at the beginning of the 21st century. Some herders dwell in huts pitched next to the houses of emigrants in town (or in parts of their houses) as guardians of their real estate capital and in many cases of their animals. It was only very recently, beginning in 2000, that the major part of the younger members of these extremely deprived families settled on lots that they bought on credit at the fringe of small urban regional or local centers.

In northern Morocco there has never existed a mobile habitat, as the concentration of torrential rains during six months demands the construction of permanent houses. In these regions, circulation has always been effectuated in the context of trans-border commerce and peasant agriculture which integrates goat keeping and forestry as well as a domestic raising of cattle for draft animals, milk consumption and the sale of calves.

Between the Atlantic Ocean and the High Atlas, between Fes/Taza in the north and Marrakech/Qalaat Sraghna in the south, 148,389 tents were dispersed across the plains and plateaus in 1931. In 2005, only some huts constituted of branches and plastic-foils remained. They either belong to very old people whose children have moved to town where they live off more or less secure revenues, or to very young households with one or two children, which, raising small flocks, try to accumulate some savings to build a house. In the censuses, these habitations are declared as permanent, but they are being relocated according to agreements with agriculturalists.

In the Middle Atlas, 27,495 tents were found in 1931. The seasonal pastoral transhumance that prevailed here was the task of entire families, of both shepherds and herd owners, until the mid-1960s. Everywhere between Ain Leuh, Ifrane, Sefrou, Aguelmous, Mrirt, Ait Isehaq, Khenifra, as well as in the piedmont lands of Oulmes, Boujaad and Oued Zem, the settlement of land-owning families in the regional urban centers or on the cultivated piedmont lands has been intensifying and accelerating over the last third of the past century.[4] Herd mobility has been reduced to daily to-and-fro-movements between the house and neighboring pastures.

On the northern slope of the High Atlas, livestock production has always been integrated in agro-pastoral systems. The small herds on these steep and narrow slopes are traditionally kept in shelters attached to houses. The smallest herds stay there throughout the year. The larger ones only spend some days in autumn or spring there. In summer, they are moved to the highland pastures where the shep-

Table 3-1: The Evolution of Nomadism in Southeast Morocco: Numbers of Nomads

Region	1931	1960	1994
Talsinnt, Bouanane	3,011	1,480	2,740
Tendrara, Bouarfa, Figuig	2,464	*3,326	3,058
Total	5,475	4,806	5,798

Source: Census data.
Remarks: *These are classified as sedentary in the census. However, this classification is probably false, for observations lead to the more adequate judgment of counting the majority of them as nomads.

herds sleep in provisional stone shelters, and they spend the winter on the piedmont plains. There, the shepherds pitch small tents, but they still consider themselves as sedentary because of their ties to their home villages where they possess houses. The 4,161 tents counted in the 1931 census for the northern slope of the High Atlas belonged to the livestock producers of the El Qsiba and Zaouiat Ech Cheikh regions at the junction of the Middle to the High Atlas. Today they are sedentary but still keep herds, like in the northwest, of about thousand and more sheep and goats.

In the eastern Moroccan region from the Mediterranean Sea south to the Sahara, until 1931 the only houses were to be found in the historic urban centers or in the areas of agrarian colonization. Nearly the entire population – some 21,507 families – lived in tents. They comprised large scale livestock producers, their shepherds and protégés, while some families were also living in tents in the vicinity of urban sites. In 1960, sedentarization had only affected the coastal front of the lower Moulouya and the plains of Angad. The populations of the southern and southeastern plateaus and plains – some 4,806 families – were still nomadic. For 1994, 5,798 families have been declared as nomads there, a number that is probably underestimated. Hence, remarkable about this region is the stability of the number of pastoral households, and thus the vitality of livestock production systems, which is paralleled by a traditional and continuous movement of emigration from the south to the north.

Within the regions in the Saharan zone[5] in southern and southwestern Morocco, the families which were declared as living in tents in 1931 practiced nomadic livestock production. Without counting the most southern populations, the number of these families declined from 7,562 in 1931 to 5,785 in 1960 and finally to 4,607 in 1994. This can be attributed to different dynamics. In the south, it reflects the importance of migration of the Ait Atta in the first half of the 20th century. In the southwest, it illustrates the bisection of the habitat between regional towns, colloquially called *dcheirat,* and the tent, which allows for the maintaining of nomadic

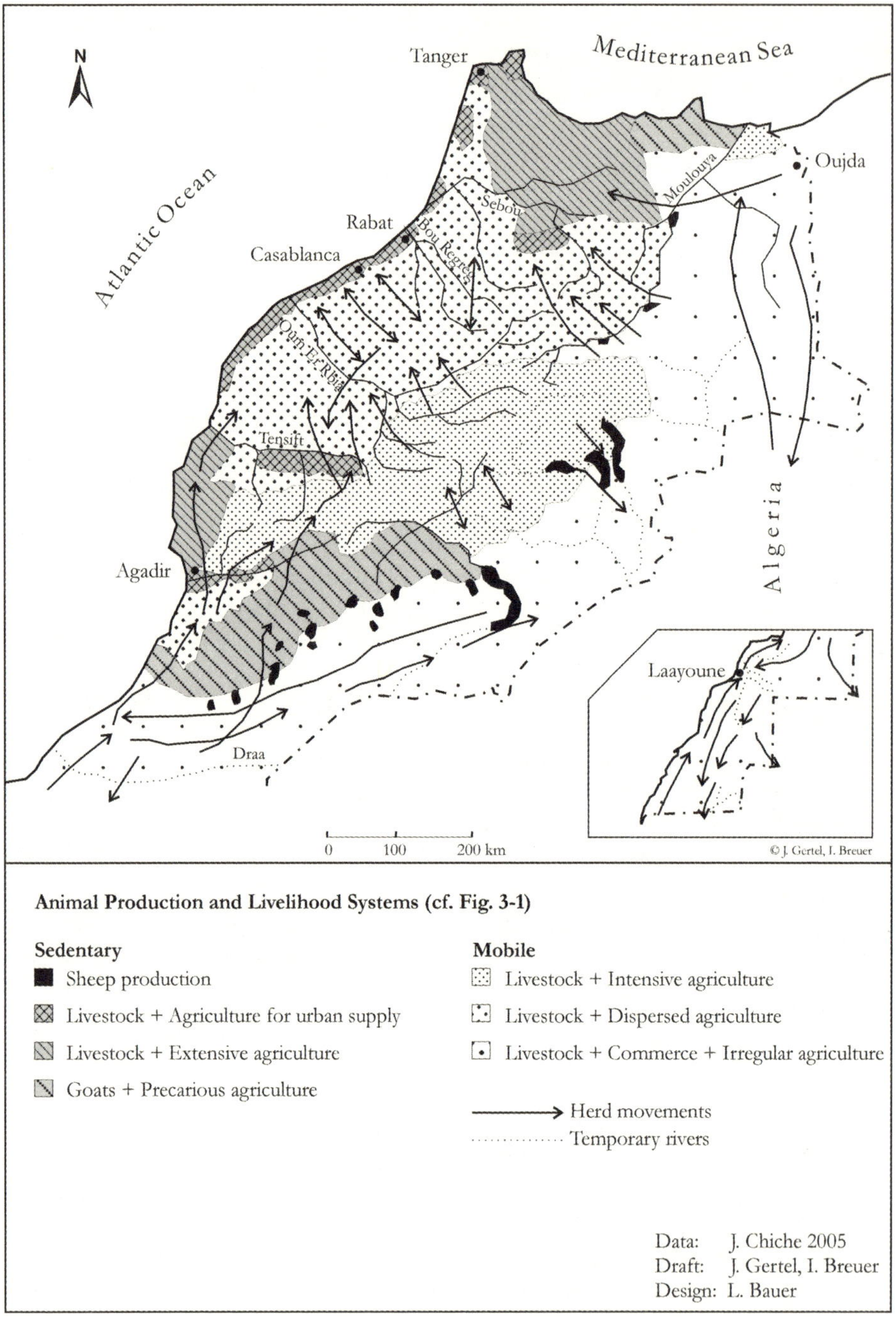

Map 3-1: Morocco 1934 – Agricultural and Pastoral Production Systems

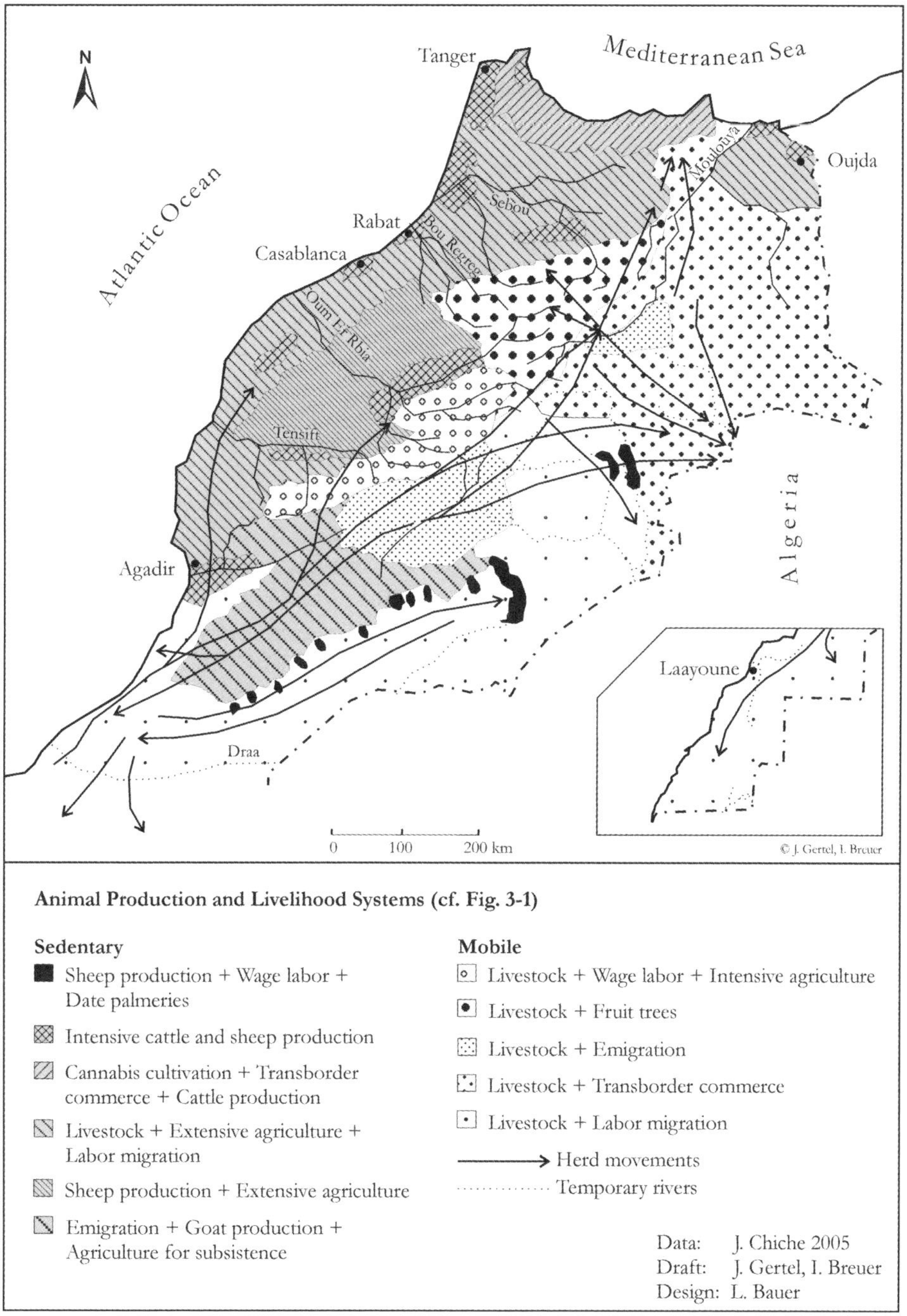

Map 3-2: Morocco 2005 – Agricultural and Pastoral Production Systems

livestock production. Finally, in the southeast, it reflects the fluctuations due to the frontier situation.

The nomadic habitat has diminished everywhere. Still, this happened unevenly across the country (Chiche 1998). The relationship which I will draw between this development and that of the future pastoral livestock production merits more attention specially in terms of livelihood strategies. I will first search for characteristics in the mobility patterns of different groups of the population. Then I will consider the issue of these groups' degree of livelihood insecurity. By doing so I will shed light on the facts that, parallel to the birth of new technical factors pressuring the nomads to settle, the objectives of mobility have multiplied, as have the determinants of each group's vulnerability.

Agricultural Modernization and Urbanization

In the course of the 20th century the most important cause responsible for the settlement of pastoralists is the restriction of land and pastures for the herds due to the expansion of cultivation land, tree plantations, urban and industrial spaces and the delimitation of forests that restrict the passage of animals.[6] Hence, across Morocco exploitation of nature and land occupation have evolved in the sense of intensification and spatial densification.

Prior to the French and Spanish Protectorates, agriculture was assigned a low priority, both socially and spatially. Permanent occupation and regular cultures were limited to several sets of circumstances, humid soils or soils on which water could be mobilized, like in the hills and mountains of the northwest (northern Jebala), or valleys of the high mountains where irrigated terraces were carved into the slopes. Regular cultivation was practiced in mountain areas exposed to Atlantic climate, in the Pre-Saharan oases, as well as in the hinterlands of cities, ports and residential sites of the *makhzen*. Further, non-irrigated cultivars were grown in the northern Atlantic plains and plateaus, and on the lowlands of the arid zone depending on climatic and political conjunctures. In general, the surfaces under cultivation also remained limited by the disposition of draft animals, laborers and seed.

The expansion of cultivated lands using cultivation techniques designed in southern Europe, forest delimitation and the creation of urban centers, have followed from a northwestern direction towards the arid southeast and southwest. The introduction of these innovations has arrived at a stage, where it seems that today the maximum technical possibilities of each place have been realized under this regime.

1934	2005

Animal Production and Livelihood Systems – Sedentary

1934

- Sheep production: Domestic sheep production of the oases.
- Livestock + Agriculture for urban supply: Production of sheep, goats, and cattle, and agriculture destined for urban supply.
- Livestock + Extensive agriculture: Production of cattle and goats, articulated with extensive agriculture destined for urban supply.
- Goats + Precarious agriculture: Production of goats and agriculture under very precarious circumstances.

2005

- Sheep production + Wage labor + Date palmeries: Income from cultivation of date palms and from wage labor in northwestern Morocco. Domestic sheep production in the oases.
- Intensive cattle and sheep production: Intensive cattle production (milk, meat), and intensive sheep meat production.
- Cannabis cultivation + Transborder commerce + Cattle production: Income from cannabis cultivation and from transborder commerce. Domestic cattle production. Recently modest resumption of small scale goat production.
- Livestock + Extensive agriculture + Labor migration: Revenue from animal sales, secondarily from cereal production. Production of sheep, goats, and cattle for savings, articulated with extensive agriculture.
- Sheep production + Extensive agriculture: Production of sheep, secondarily of cattle, permitting the quick mobilization of capital, articulated with extensive agriculture.
- Emigration + Goat production + Agriculture for subsistence: Income from commercial activities connected to emigration. Negligible amounts of goat production and agriculture for household consumption.

Animal Production and Livelihood Systems – Mobile

1934

- Livestock + Intensive agriculture: Production of sheep, goats, and cattle (intensive in winter, extensive in summer), articulated with intensive agriculture in the narrow valleys.
- Livestock + Dispersed agriculture: Production of sheep, secondarily of goats and cattle, articulated with dispersed and more or less regular agriculture.
- Livestock + Commerce + Irregular agriculture: Production of sheep, goats, and dromedaries, articulated with commercial activities and irregular agriculture in the basins.

2005

- Livestock + Wage labor + Intensive agriculture: Income from seasonal wage labor in the construction sector, from trade with agricultural products, and from production of sheep, goat, and cattle (intensive in winter, extensive in summer), articulated with intensive agriculture in the narrow valleys.
- Livestock + Fruit trees: Income from animal sales and from cultivation of fruit trees, plus large sheep and cattle herds for slaughtering and for speculation.
- Livestock + Emigration: Income from commercial activities during emigration, and from sheep and goat production. Ongoing expansion into other regions.
- Livestock + Transborder commerce: Income from transborder commerce and from animal sales. Nomadic sheep production, recent sedentarization.
- Livestock + Labor migration: Income from commercial activities and animal sales. Nomadic sheep, goat, and dromedary production.

Fig. 3-1: Annotations for Map 3-1 and Map 3-2

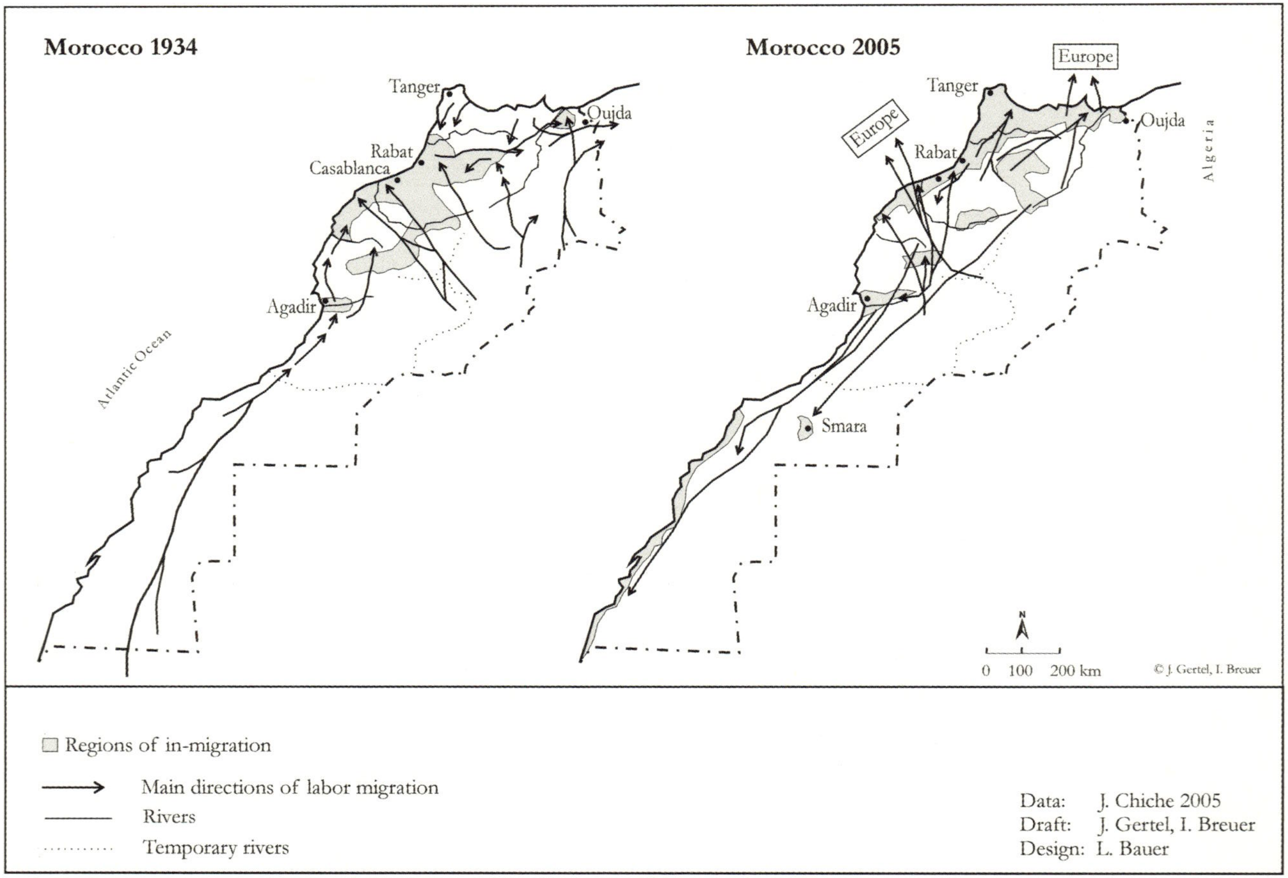

Map 3-3: Morocco – Labor Migration (1934, 2005)

The best results are achieved in agricultural regions where availability of water and soil coincide with easy transport of products and means of production, as well as strong connections to markets. Due to ongoing expansion and concentration processes these agricultural regions are currently overexploited.

Cultivation on the Atlantic plains, on the plateaus of Doukkala and Chaouia and on the plateau of Meknes and Fes has been followed by a progressive decline in herd mobility from 1912 up to the middle of the century. This often meant the separation of families into two branches: mobile and sedentary. Sedentarization began early among the Zaer and Zemmour, and later – between 1956 and 1960 – in the arid Atlantic plains among the Rehamna, Beni Meskine, and the Sraghna, and, finally, in a movement fulfilled at the end of the 1970s, on the western fringes of the Middle Atlas between El Hammam and Khenifra and on the plateaus of Boujaad and Oued Zem. The reduction of frequency and radius of the winter descent of herds from the northern slopes of the High Atlas and the high Chaouia towards the plateaus and plains of the Beni Meskine, Rehamna, and Sraghna are consequences of farmland expansion. In other semi-arid and arid regions the habitat of humans and animals has seen a progressive immobilization. Still, pastoral livestock production has not lost its pre-eminent position among the peoples' activities. Despite the cessation of transhumance, the herds remain very large. The reduction of herd mobility began at the end of the 1950s in the regions of Aguelmous and Oulmes, parallel to the expansion of cultivation at the western fringe of the neighboring Middle Atlas and in the mountainous terrain between the Middle and High Atlas.

In the eastern Moroccan steppes the pace of sedentarization and agricultural expansion has been synchronous, starting in the northern plains after 1910, and experiencing a renaissance in the 1970s in parts of the regions of Taourirt, Tafrata and Boumia. This latter wave of sedentarization was induced by the introduction of tractors and was motivated by a defensive attitude towards the menaces of restricted land access caused by the creation of programs for pastoral amelioration. Since 1990 sedentarization continues with the planting of olive trees and the abandoning of extensive livestock production in favor of the profitable intensive cattle breeding around Guercif and even more discretely with the construction of stables in the Dahra, the central part of the eastern Moroccan steppes (cf. Tag, this vol.). In the Dahra, sedentarization started immediately after 2000, when pastoral families constructed houses directly on the pastures. But on these arid soils, sedentarization is not accompanied by a large-scale expansion of arable land.

In the northern part of the Middle Atlas tablelands, a gradual retreat of herd movements to the high-altitude pastures occurred from 1912 on. It followed the introduction of cultivation in the piedmonts by large livestock producers who were encouraged by the preparation of infrastructure for cultivation. Nomadic livestock production was further pursued in a system where herds were entrusted to shepherds, who in turn employed auxiliaries, while the landlords settled down on farms or in the regional towns with their families. Yet, the reduction of herd mobility has been laggard and unspectacular till the early 1980s. Permanent sheep stalls erected at high altitude on small plots cultivated with barley are increasingly important, and herd mobility could be further reduced due to the possibility of motorized transport of supplementary fodder.

Many pastoralists with large flocks continue to be mobile and even progressively enlarge their field of activity. Traditionally this is the case in the upper Moulouya and the Atlantic part of the Saharan zone, and recently since the late 1990s on the southern slope of the central High Atlas, outside of Ouarzazate. This evolution is certainly permitted by technical factors, notably the motorized transport of animals, water and fodder. Meanwhile, it seems evident that these livestock producers advance to the conquest of new spaces.[7] The orientation of the nomads from Ouarzazate towards southeastern and southwestern Morocco – in an inverse direction compared to that which has historically guided the nomads – also confirms the emergence of new centers for agricultural and urban development, which progressively attract investment and population.

In the 20th century human mobility has rapidly gained in importance and changed its character. In former times, the principal routes were those of camel drivers, shippers, merchants, and livestock producers. They accomplished the transport and trade of merchandise between Sub-Saharan Africa and the ports of the Atlantic and the Mediterranean coast, in passing by the oases of the Saharan fringes, then by posts guarded by garrisons and *guich* groups. Concomitantly, in the course of centuries, Moroccan pastoralists pursued a gradual and continuous movement from the southwest towards the plains of Marrakech and Abda and the hills of Chiadma and Haha, and from the south towards the northwest across the Middle Atlas.

The conjunction of economic and institutional initiatives of the new political order led to a boom of coastal and internal towns. This was based on the implantation of industry and administration, the agrarian colonization of the plains and their hinterland, the development of vegetable cultivation in the plains of the middle Atlantic coast, the creation of mining centers at Khouribga and at the south of Oujda, and finally the introduction of a legislative system favoring individual choices. Population has increased in the urban areas, as well as in the agricultural regions of the Atlantic plains and plateaus. On the northern front of the High Atlas, people have gradually descended from the mountains towards the piedmonts, where replications of their villages of origin have been created, while the original villages have practically been abandoned.

In summary, the regions of intensive agriculture and the cities have drained the majority of migrants from rural areas, while pastoral mobility has little by little withdrawn to the south and southwest of the country since the 1930s. The movement of nomads across the Middle Atlas has not ceased, but it is restrained by the expansion of fruit plantations. From the beginning of the 1980s, the saturation of the Atlantic plains with cereal lands and expanding agglomerations concomitant with the availability of new techniques increasing the productivity of irrigated cultures destined for commercialization have multiplied the number of households engaging both in agriculture and speculative animal production. This happened all over Morocco, but notably in the hills and mountains of the north and in the alluvial plains of the southwest. Subsequently, these regions have become poles of attraction for seasonal migrants, who nevertheless remain attached to their place of origin.

Transformations in Livestock Production

A review of the data concerning the national stocks of sheep, goats and cattle (Table 3-2)[8] well illustrates that the declining number of tent dwellers and the decrease in herd mobility did not entail a corresponding decrease in animal numbers, as was witnessed in many other countries. Up to the 1960s cattle have held a notable place among the draft animals, especially in hilly and mountainous regions. During the first half of the century, their numbers have constantly been in the order of 2 million. From the early 1970s to the mid-1990s, the encouragement of milk and meat production had the effect of augmenting the national stock, which then varied between 3 and 4 million head. At the same time, an intensive animal husbandry of imported cattle breeds was implanted. Nevertheless, the traditional cattle stock, which is a less productive, but more resistant breed, remained important. Since the BSE-crisis of the 1990s, and most likely due to 20 years of experience in economic management of intensive livestock production, the number of cattle has decreased and remains around 2.5 million head. At the regional level, the most outstanding increase in cattle numbers has taken place in the arid zone, where great irrigation projects have been developed. In the mountains, the number of cattle has sensibly rested unchanged. In the plains, the motorization of agriculture and the spread of stable keeping has resulted in a slight decrease of stock.

The number of sheep follows a general tendency towards augmentation with mitigation of losses in case of bio-climatic crises. From 1940 to 1960, the herd has fluctuated between 12 and 14 million head, after a regular growth bound to the progress of prophylaxis. Since the introduction of supplementary fodder in the 1960s and 1970s, the size of the national stock varies between 14 and 16 million head with a stabilization around 16 million since the mid-1990s. In the early 1950s, the number of goats has increased from less than 2 million head to nearly 10 million, but till the end of the 1970s it dropped to between 5 and 6 million. This was due to pasture restrictions imposed by the reinforcement of forestry regulation after Morocco's independence. After that, the general decline continued, but the number of goats experienced a slight upswing from 1995 to 2000, and a phase of stabilization till 2004.

Table 3-2: Number of Cattle, Sheep, and Goats in 1932, 1958, and 2004

	– Cattle –			– Sheep –			– Goats –		
Region	1932	1958	2004	1932	1958	2004	1932	1958	2004
Oriental	30,671	33,502	50,900	494,567	809,515	1,471,200	195,368	290,598	337,300
Nador	?	76,511	80,700	?	254,328	631,100	?	221,283	124,700
Fes and Taza	272,826	296,968	253,700	1,171,398	1,587,959	1,954,900	801,231	857,261	549,600
Meknes	32,910	229,594	123,800	655,301	1,792,627	1,607,500	194,485	1,062,012	246,000
Tetouan	>60,877	216,257	274,800	>167,322	357,643	460,300	>129,414	526,247	572,300
Rabat north	203,552	273,743	164,000	576,840	896,308	903,300	79,006	130,851	21,600
Rabat south	201,495	221,186	137,000	473,290	857,739	612,600	229,020	234,290	140,600
Casablanca	287,202	260,407	249,000	816,216	908,159	2,058,200	124,375	37,077	87,000
Doukkala	215,732	271,900	236,600	311,114	416,557	587,400	15,894	4,739	4,600
Tadla	72,840	183,706	168,100	459,030	1,389,596	1,049,400	172,539	519,579	547,200
Marrakech	208,123	371,799	306,300	725,129	3,313,741	1,762,500	536,639	559,932	511,800
Marrakech coast	162,133	231,242	185,400	487,910	1,237,741	1,420,200	361,639	522,732	432,200
Agadir	78,722	216,161	339,100	>274,902	>568,049	1,839,900	>228,631	>1,082,459	1,782,700
Total	>1,827,083	2,883,029	2,569,400	>6,613,009	>14,389,962	16,358,500	>3,068,241	>6,049,060	5,356,600

Source: Ministère du Plan: Annuaires statistiques du Maroc, different years.

The statistics thus reveal that after each crisis, the number of goats fails to reach its preceding level, while that of sheep not only reaches its former level, but even surpasses it. Hence, the role most producers attribute to livestock differs depending on the kind of animal. The goats, in general herded very extensively and in ways that make them comparatively resistant to climatic risks, serve to cover the running costs and are used for the immediate reconstitution of stock after crisis. Sheep, in contrast, are considered of greater value and are a more profitable product. They are also used as savings to be mobilized at the time when heavy investments are necessary, for example, for the purchase of land or equipment. Consequently, pastoralists devote a considerable financial effort to sheep raising. Finally, cattle are raised as a purely commercial product, which follows a continuous circulation. The diminution and stabilization at a number inferior to that of the 1940s and 1950s result from the development of intensive livestock production systems with a smaller number of animals.

The evolution of stock is also strongly determined by the introduction of new livestock production techniques. Prophylaxis against diseases and parasites has been performed from the beginning of the 20th century. Subsequently, it has gained momentum, especially for sheep and cattle. Its effects on the stability of herd sizes are represented in Table 3-2. Still, it is rather recently, in the course of the last decade, that prophylaxis entered the norms and everyday practices of livestock producers.

For ages before Morocco's independence, the cross-breeding of sheep and goats from different parts of the country with those of neighboring countries, like Mauritania and Algeria for sheep, and Spain for goats, had been pursued sporadically. However, the systematic import of foreign sheep and cattle breeds, especially for wool and meat production, was limited essentially to the hinterland of Casablanca and Rabat.

Since the 1960s, but above all since 1975, the state has undertaken strongly organized support programs for the intensification of animal husbandry, including the conception and realization of a development plan for the raising of dairy cows, and a feedlot plan for he-goats and rams.[9] These programs integrated measures like import of selected animal breeds, subsidies on animal fodder, the creation of livestock production cooperatives and professional associations, and construction of stables and centers for milk collection. Furthermore, the plans fostered the installation of an economic system connecting cultures, animal husbandry and agro-alimentary industry in the grand irrigational areas. Later, in the 1990s, the intensification of dairy goat raising was launched with impetus from different NGOs.

Parallel to this, from 1992 onward, the public services have taken the initiative to reduce the effects of climatic crises on pastoral and semi-intensive livestock production. This was done by amelioration of small ruminants' resistance and by placing subsidized external fodder within livestock producers' reach. Pastoralists in all regions adopted and spread this practice after 1970 (cf. Rachik, this vol.). Yet, it must be emphasized that the degree of supplementary feeding differs crucially, depending on the kind of animal, the region's vegetative productivity, the livestock producer's financial assets, and even the envisioned level of production.[10]

Some livestock producers use trucks to transport animals between the pastures in different regions. This is especially true for pastoralists from the arid zone (cf. Breuer, this vol., High Atlas). Herds from the High Atlas or the high Chaouia, around Settat, may be sent to neighboring fallows by truck. Until the early 1990s, the relocation of herds was in general accomplished on foot, except some inconsequential initiatives, strikingly the sending of animals from the eastern Moroccan steppes to the region of Taza by train during the droughts of 1945 and 1946. The effects of the motorization of pastoralism on vegetation dynamics have hardly been studied yet. Considering the concentration of herds in certain regions, due to the regrouping of large herds in convoys, it would be an interesting issue for further reflection.

Changes in the Consumption of Livestock Products

Another change affecting livestock production is the transformation of consumption and demand for animal products. In total, the purchasing power and nutritional level of the population's majority are relatively low. The demand for animal products essentially concentrates in the cities. The only consumers eating meat on a daily basis are the inhabitants of wealthy urban areas (cf. Khalil, this vol.). Families on more modest budgets indulge in poultry or fish rather than red meat. Their consumption of animal products is generally infrequent and accords to what their budget allows. Consumption of butter and milk is more widespread among this group, as it is among work migrants employed on construction sites. In the country, a social partition of the same type can be observed, while the total level of meat consumption remains low. The evolution of the products' origin demonstrates that the market integration of rural areas is well underway (cf. Breuer, this vol., Pre-Sahara). In Morocco, the demand for milk and dairy products is met by local cattle keeping and imports. The increased demand in the 1970s has been paralleled by a production rise bound to the 'milk plan' from 1975. Despite these facts, consumption remains very low at an annual equivalent of 37.75 liters of fresh milk per person in 2001, with a considerable discrepancy between urban (53 l) and rural areas (18 l). Consumption has grown in the urban areas only, where increasing supply of treated and packaged products has facilitated access. In contrast, consumption dropped from 20 to 18 liters in rural areas. At the same time, the conquest of the rural markets, underway since more than 15 years, was vehemently pursued. As early as 1985, for example, 24% of milk consumption came from purchase, compared to 33% in 2001.[11]

Above all, the consumption of butter and margarine has risen. In rural areas it even has surpassed the consumption of clarified butter. This increase is due to the introduction of treated imported butters and margarines, which are as frequently consumed in the country as in the cities.

One prominent character of the nutritional table in Morocco is the dominance of fish. The annual fish consumption lies at 11.64 kg per person in urban areas and at 5.67 kg in the country. In urban areas, fish is followed by industrial poultry (9.19 kg per person and year),[12] and by beef (6.5 kg), mutton (4.73 kg), and goat (1.15 kg). In rural areas, this ranking is the same, with lower figures for each item.[13] Meanwhile, statistics and observations indicate both that the *Aid el-Kabir*-celebration remains a principal occasion of meat consumption for most Moroccans, and that different qualities of red meat and fish are offered to different kinds of customers on the *souks*.

The consolidation of the state has progressively contributed to an urban nexus. Urban growth and the creation of small towns have been fostered by Moroccan public services, especially in the mid-1970s and since the beginning of the 1990s. Small centers with considerable service functions have been born throughout the country.[14] Thus, if we analyze consumption patterns, we have to take into account the enlarged infusion of transformed imported products, and also the fact that the population is increasingly concentrating in small service centers while maintaining a rural lifestyle. Whether in towns or in rural areas, these results confirm that for one part, the consumption patterns of the wealthiest groups have improved throughout the second half of the 20th century, in particular within the last two decades, while the alimentary and nutritional level of the majority stagnates, if it does not sink. These changes are mainly caused by market integration and introduction of market produce.

Livestock Production and Livelihood Systems: A Regional Overview

Within these trends and constraints there is nonetheless enormous variation across Morocco in the experiences and practices of pastoralists. The key to these variations is rooted in the livelihood strategies followed by different groups. What follows is a region-by-region review of livestock production systems. It aims at assessing the economic importance of pastoralism in rural Morocco and at identifying the divergent manners in which pastoralism relates to other economic activities and income sources. The regions which will be discussed are the Anti-Atlas, the southern oases, the northern Atlantic plains, the northern Moroccan mountain areas, the northwestern Middle Atlas, the northern slope of the High Atlas, the fringes of the Middle Atlas, the steppes and the Pre-Saharan regions, the eastern Moroccan steppes, the southern slope of the eastern High Atlas, the central and eastern Sahara, the southern slopes of the central High Atlas and the upper Moulouya. Finally, the urban peripheries and the fringes of large irrigation sectors will be treated as one last regional unit.

The Anti-Atlas: Livestock Production Remains Limited Due to Scarcity of Water Resources and Emigration

Virtually all families in the Anti-Atlas live off commercial revenues and salaries sent home by male family members working in the food retail sector in Moroccan or European cities. The climate's aridity, the absence of general access to ground- and surface water for domestic and productive usage, the extreme poverty of soils, and the tradition of emigration resulting from this, are at the origin of the predominantly symbolic character of agriculture and raising of some goats. It is only on the western more humid mountain fringe that families raise strictly stable-held cattle. The households that stay in the region are reduced to aged men, women, and children. The men who succeed in assuring a regular revenue in their place of immigration choose to take their families with them, and they guard only sentimental ties with their home community. It is only with very advanced age that they return to live in the countryside, where they pursue the mode of consumption acquired in town. Those who create an enterprise or find a regular job, and still cannot guarantee a satisfying life for their families in town, leave their families in the village and return periodically to provide them with water and consumer goods. Recently, the region has however become a pasture zone for herds of dromedaries coming from the southwest and heading towards the plains and piedmonts of the northern slope of the High Atlas. They circulate without constraint or contestation all across the mountains, since the locals are not interested in the area's natural resources.

The Southern Oases: Livestock Production Plays a New Role as an Instrument of Technical and Social Intervention

In the oases of Tafilalt, of the middle Draa valley and of the Jebel Bani, as well as in the valleys of the southern slope of the central High Atlas, the date palm cultivators live from selling dates and henna and from working all over the country, essentially as masons or as specialists for well digging. They practice domestic livestock production of one to three goats. In the Tafilalt, the agrarian authorities of the ORMVAT[15] have initiated the keeping of dairy cows, and strongly support it both financially and technically. The animals, sheep of the *Demman* and goats of the *Ticherguiat* breeds, are permanently kept in the houses' hallways, where they are fed side-products of cultures, weeds, lucerne, crushed date seeds, straw, bran, and cereals. Since the late 1980s, the sale to slaughterhouses, and thus market integration, has gained importance. This was due to aid programs that encouraged the constitution of large intensive enterprises with up to 400 and even 1,000 ewes, combined with large-scale market promotion. But the radius in which oasis meat was marketed could not be expanded. These projects were nevertheless viewed as a great success by development agencies. Sometimes they served as nuclei for small domestic herds distributed to poor women by NGOs, or as an instrument for the augmentation of fertility of local livestock. Hence, some speculators specialize in the purchase of animals for reproduction and donations while the majority of families pursue small domestic animal husbandry.

The Northern Atlantic Plains: Cultures Dominate, and Livestock Production Still Plays the Role of Liquid Capital Among the Most Needy

In the plains and lower northern Atlantic plateaus,[16] in the western High Atlas, in the lands of Chiadma and Haha, in the hills of the Sidi Qacem region, on the plateaus around Fes and Meknes, and in the Taza region, livestock production is combined with agriculture, services, and other salaried revenues. This is true on all social levels: The largest herd owners, for example, have diversified their branches of social and economic investment and can, according to the general economic situation, build up large herds or sell them out. But, on the other end of the social scale, even the occasional seasonal workers possess animals.[17] Each family has a sheep herd, albeit rarely large. Some even own cows, but generally very few. It is among these latter ones and the middle-scale agriculturalists that herd alimentation is based on agricultural products, on fallows and roadside vegetation. The guardianship is entrusted to the families' youngest children, to elders, to distant relatives, or to women without resources who the family head may take as second wife to accomplish that task. The herd may also be shepherded by a daughter, who is then married as late as possible, contrary to local custom and thus sacrificing her future. Alternatively the herd may be entrusted to an adolescent from the neighborhood, who receives salary, food and accommodation. The animals are sold on the local *souks* to small *maquignons* who in turn sell to butchers or grand speculators. According to the relationship between capital and family size, the lion's share of revenues may come from agriculture, livestock production, industrial or service activity, or from a salaried employment.[18] Livestock production dominates the revenues of two groups: the poorest, and the rich livestock owners. For the poorest, the sale of an animal, very often in bad condition and consequently at a very low price, constitutes a major income source. By means of their few animals, they constitute a permanent capital which they can rely on in times of pressing need. It may also be the case that their stock serves as the basis for an investment, should a good opportunity present itself. The large scale livestock producers, on the other hand, pursue the same strategies as the pastoralists of the northwestern Middle Atlas, but with the handicap of restrained land access, as almost all land is private property.

The Northern Moroccan Mountain Areas: Livestock Production is Losing its Economic and Technical Function

In the northern Moroccan mountains, animal husbandry of small resistant cattle remains important for the agricultural soil preparation, the daily consumption of a little milk, and the sale of calves all 12 to 18 months. Almost all families keep cattle. Goats, in contrast, which for centuries had been part of an agro-pastoral system combining seasonal cultivation on forest spaces and livestock production for fertilization, thinning and alimentation, have been abandoned in the mid-1980s. This was due to the growth of revenues from the cannabis cultivation, which had begun to stabilize through market conjuncture, mineral fertilizers and motorized transport. Since 2000, the launch of aid programs for economic and social intervention by public services and NGOs has stimulated renewed interest in the amelioration of goat keeping. There are now only very small herds raised with little intensity. Traditionally guarded in small enclosures, they are fed by the women who bring them branches, and shepherded by children in the morning and the afternoon. It seems very likely that this transformation towards domestic livestock production was performed to the benefit of a greater monetary liquidity, in the context of nationwide speculation on goats and *Demman* sheep and tactics of maintaining good relationships with organizations devoted to rural development.

The Northern Slope of the High Atlas: Livestock Production is Intensive in Winter and Extensive in Summer

The population of the northern slope of the High Atlas can be divided into three groups: First, many families live off salaried work in the construction sector, favored by the boom of urbanization across the whole country. Second, merchants collect local products, animals, vegetables, nuts, and other fruits. Third, retail traders of daily consumer goods, like flour, oil, sugar, tea, and gas, supply the masons' families on credit during the whole duration of the masons' absence. Some of them enlarge their farm by leasing land and water from these seasonal migrants. Most families live according to a rotational system of different age groups, in which the men of working age seek employment outside the region, and in which the aged individuals, the women, and the children reside permanently in the villages and take care of the fields and the animals, which are almost always family property. Families which do not possess any animals are very rare. The largest herds, however, only exceptionally exceed 50 head. Generally the herds are small. Their offspring is sold on local *souks*. The small village *maquignons* sell the small ruminants to the *souks'* butchers, while calves are bought for fattening by butchers from the towns in the plains. In winter, livestock production is conducted intensively, and the animals are kept in basement stables of the houses and nourished by branches cut in the woods or by hay mown throughout the year on farmland parcels or collective lands. They are led outside by a family member when the weather is neither too cold nor too rainy. In summer, the animals mount to the high altitudinal pastures, in proximity to the village, opened at a date decided upon in local agreement after its closure during the thawing period and for vegetative regeneration. The first day after the opening, all herd owners and their families climb up to the summer pasture, but only some shepherds to which each one confines his few animals sojourn there throughout the entire season. Now this system seems in radical rupture with the traditional one. A short while ago, revenues were based on agriculture, but today they are built on the salaried work in market-integrated regions. This means the introduction of a supplementary income source to the economic portfolio of households (cf. Gertel, this vol.). Traditionally, the relationships to the exterior depended on the sale of local products in the towns during winter. This was performed by men who descended individually from their villages. Today, each family calculates with scrutiny the part of salaried work, the sale of agricultural produce and that of animals in their budget. Yet, the living standard, the level of consumption and the livelihood security remain low for the whole of the population.

The Northwestern Middle Atlas: Pastoral Livestock Production is Integrated into Modern Markets

In the western Middle Atlas, where the regional climate is cold and humid, livestock production systems have progressively specialized in market production. All the whilst, production remains pastoral, though it is stabilized by feeding, which may be more or less regular and abundant. The sheep of the *Timahdite* breed and cattle of different types are sold on the *souks*. The valorization of this meat by urban consumers endowed with strong purchasing power, besides being reputed for its high quality, draws as much from the real superiority of taste, as it draws from the ecological image bound to this breed. It also benefits from the fact that an important part of the populations of the cities in which this meat is consumed, like Meknes and Fes, is constituted by former inhabitants of the Middle Atlas. Livestock production is thus well conducted, technically and economically. Most herds are kept according to a controlled pastoral system. The animals are brought to the high altitude pastures as long as resources are sufficient. They only receive a regular and timely minimized extra fodder, especially of barley, bran and straw, during transitional periods, bound to the seasonal cycle. Alternatively they may receive fodder input during periods of unforeseen occurrence and severity. Some of the male animals are fattened, the bulls all year long, the sheep essentially before *Aid el-Kabir*. Stall keeping is however reduced to a temporal minimum, as long as it is thought possible to fatten the animals on grass. However, only wealthy herd owners, who constitute a minority in the region, dispose of pastoral resources and financial means permitting a good maintenance. These families of large livestock producers, firmly rooted in the region, are divided into independent households, each living in a different social and natural environment – cities, regional towns or commercial farms – where they conduct different activities. The majority of these activities are highly integrated and articulated into the regional and national business world. The pastures of high altitude are more voluntarily confined to shepherds (cf. Djoudi, this vol.). Other families of herd-owning pastoralists, some of which possess only few animals, and the majority of which are established in the region for only a relatively short time, live under conditions of mediocre comfort, often near the sheep stalls, and sometimes in tents installed on the high altitude pastures. The vitality of this mode of livestock production that has succeeded in acquiring a place on the modern market is reflected in the fact that the region belongs to those attracting permanent migrants.

The Fringes of the Middle Atlas: Livestock Production is Made the Object of Speculation

On all the western surrounding and the eastern fringe of the northern Middle Atlas, namely the massif of Oulmes, the plateaus and plains of Aguelmous, Khouribga and Rehamna, the mountains of El Qsiba and Zaouiat ech Cheikh and Jebel Bou Iblane, the primary source of revenue is pastoral livestock production. Two forms of sheep production have crystallized: one very extensive, the other intensive in stables with fattening before sale. Besides these, there exists a less significant husbandry of bulls for slaughter. The stock is owned by a handful of families only, which in their entirety are patriarchic. These families are made up of local and regional elites, and their members commute between different habitations, normally two to four, divided between the pastures and local and regional towns. Each of these families owns between two and ten sheep herds. These are herded by salaried shepherds in the forest, on fallows, on grass lands or in low ligneous vegetation. Those who raise bulls or dairy cows employ one or two salaried shepherds. The fodder distributed to the animals is either purchased or provided by the family's agricultural production. The men of herd-owning families devote themselves to the economic and commercial management of agriculture and animal husbandry. The animals are sold on the *souks* in the frame of social networks that are tactically maintained, and which are founded on regional origin and affiliations. In regions very distant from large urban centers, milk is sold to cafes of neighboring small and medium towns.

Employed as shepherds, some families profit from low but more or less regular revenues. The population's structure evinces this profession's contribution to the permanence of large families with many children. It also illustrates the appreciation of the shepherding profession as precarious and rendering uncertain and irregular revenues. Still, these facts are perceived as less bothering than the rudeness and isolation of a life in a hut in the midst of the forest. In effect, the shepherds are generally young, and as their children reach the age of exercising temporary salaried work, they seem to orient themselves towards a diversification of their activities, prominently the emigration to the periphery of an urban center. On this basis, those which succeed may be able to acquire a tiny capital in the form of a small patch of land and some animals. The herd owners are thus rural entrepreneurs with a strong integration into the national market. In effect, their strategies are centered around the transformation of forest into agricultural land, speculation on stock, and the more or less legal sale of wood. This is especially so in the eastern Middle Atlas, where there are still considerable forest areas. Their strategies are based on maintaining their capital at least costs, on keeping track of the meat and livestock market and finally on mastering the political contexts in which their economic actions are embedded. In the regions of Khouribga and Rehamna, the speculation on extensive livestock is slightly different. The ligneous vegetation is limited to occasional scrub of two kinds, *Ziziphus lotus* and *Sumac pentaphyllum*. The principal exploitation of natural resources, which is phosphate mining, only affects rural Moroccans to the extent that it limits pasture spaces and perturbs the growth of vertebrates, due to the contamination with fluorides. The economy relies, to different degrees, on raising sheep of the *Sardi* breed. This activity is individually combined with revenues from other out-of-district activities like fishing, commerce and salaried work. The *Sardi* sheep possesses a privileged position on the market, due to its large stature. They are crossbred with animals from other regions and are used for ritual sacrifice in religious and family celebrations. They are sold on the *souks* in all of the country's south, down to Guelmim and in the central High Atlas. Results and continuity of these transactions vary according to local conditions and the livestock producers' capacity to secure sufficient fodder. Favored by the upswing of its traditional and cultural valorization, the breed is progressively distributed across a large part of the country, where it leaves its impregnation on local types.

The Steppes and the Pre-Saharan Regions: Livestock Production Remains Nomadic with Variations

In the arid zone, mobile livestock production is generally performed by motorized transport of animals, fodder and water between different pastures. The importance external fodder gains here, is more bound to the introduction of cattle and the stabilization of ovine stock than to the augmentation of distributed rations to sheep.[19] In general, the practice to use extra fodder varies from region to region, each of which is marked by very particular strategies and territorial and technical modalities:

The Oriental Steppes. In the Oriental steppes, the most ancient documents attest to the fact that for centuries animal husbandry has been nomadic, albeit closely associated with other activities and revenue sources, like commerce, collection of tributes, and ownership of palm trees. At the same time, however, some families of herd owners with medium-sized herds or of shepherds ruined by a calamitous accident, settled around towns. For more than one century, this frontier zone has been the object of assiduous interventions on the part of the public services. Each wave of sedentarization was thus tied to the beginning of one of the programs of agricultural and pastoral improvement, which have continually followed one after the other since 1912. The most recent transformation affects the central part of the high plateaus and high plains (cf. Mahdi, this vol.). It is bound to the delimitation of territories at the level of small social groups, the emergence of motorized transport, and the vivacity of the transfrontier commerce with Algeria. Livestock raising remains pastoral, and reductions in the radius of transhumance are generally due to bioclimatic conjunctures. Herds may be sent to the stubble fields of the high Gharb, the scarcely occupied lands in the southern part of Nador province, and to the pastures of the southern part of the steppes between Tendrara and Figuig, where mobility still rules. These livestock producers are proud of the purity of their stock, the *Beni Guil* breed, which is technically and culturally highly esteemed. Paradoxically, they adopted practices that they themselves denounce as damaging their pastoral economy. In effect, the introduction and proliferation of the *Oulad Jellal* breed from Algeria has shaken the principles of the region's livestock production in the course of the past decades and has effected an intensification (cf. Khalil, this vol.).

The Southern Slope of the Eastern High Atlas. The large scale pastoralists of the Ait Hdiddou and Ait Morghad tribes of the southern slope of the eastern High Atlas always practice a transhumance between the mountain highlands and the plains of its southern piedmont, around Er Rachidia, according to a traditional mode of livestock raising. Yet in this region, pastoralists constitute a minority amongst families living from weak revenues and conducting an oasis-type animal husbandry.

The Central and Eastern Sahara. In the Saharan part of the arid zone, from the Atlantic ocean to Figuig, in passing by the strips of the lower Draa, the Bani, the Draa's curve, and the surroundings of the Tafilalt, husbandry of camels, goats, and sheep is pursued in ways unadulterated for centuries, always associated with a formerly trans-Saharan, now transfrontier commerce. Automobile transport has been crucial to livestock production and commerce, because after four or five decades where it adversely affected the local economy, it has effected a renaissance of these two activities. In addition, the rebirth of frontier conflicts at the end of the 20th century and the economic evolution have entailed different transformations with rather variable implications for the practice of nomadic pastoralism. The herd movements between the east and the west are continually pursued (cf. Werner, this vol.). The peregrinations towards the south, however, have been strongly limited. This was due to several factors, such as political conjuncture, the preparation of new arable land for intense cultivation, the increasing abandonment of overexploited land further north, and more recently by the constitution of biodiversity reserves. These limitations have, for some time, served to legitimize the reinforcement of nomadization towards the north, which nowadays extends to the Souss valley and up to Marrakech and Safi. One consequence is that in the Souss, for example, power struggles between the pastoralists, typically large herd owners, the users of Argan forests, and the owners of intensive farms, are often carried out in violent conflicts (cf. Turner, this vol.).

The Southern Slopes of the Central High Atlas and the Upper Moulouya. In the area of Ouarzazate, on the southern slope of the central High Atlas, the social and economic life is organized according to a system of patriarchal enterprise dictated by a head of household who delegates an activity and a place of residence to each of his brothers, sons or nephews. This may be a public function in town, salaried work, a service activity, for instance in the transport or commercial sectors, agriculture in one of the oases, or pastoral livestock production. Those who include livestock production among their pluriactivity still constitute a minority. Livestock producers' families are often large (cf. Breuer, this vol., High Atlas), so that the patriarch disposes of sufficient relatives to oblige some of them to lead a simple life in the mountain caves, in the stone-built sheep stalls and in tents on the lowlands, only encountering sedentary people and the relatively great comforts of the oasis on very rare occasions. Since the late 1990s, the herds, traditionally moving between the line of the mountain crests and the lowlands, are progressively sent by truck towards vaster pastures, marked by the more dynamic *souks,* in southwestern or eastern Morocco. These new places of transhumance are quickly becoming the forerunners of the expansion of these family enterprises, prominently in agriculture and commerce.

For centuries the purely nomadic pastoralists of the upper Moulouya, around Qsabi, Midelt, Missour, and Talsinnt, have been living uniquely from pastoral livestock production. Only in the last few years, their animal husbandry has begun to gradually open to the market. Herd mobility generally limits itself to a radius of 35 to 50 km, which is partly due to communal partition. But in the long run, society's dynamics are strongly determined by the movement of families or individuals towards the northwest. Some of them settle in regions around the southeastern Middle Atlas, like Guigou, Boulmane, or Zeida, while others wander further to settle between Timahdite and Sefrou. We thus can conclude that these two groups of entrepreneurial livestock producers adopt the strategy of leaving their places of origin, which have become ecologically, economically and socially overcrowded. They conquer new spaces, better befitting their respective objectives – like diversification and intensification in the case of the Ouarzazate livestock producers, or the enlargement of pastures and the reinforcement of their animal husbandry in the case of the groups from the upper Moulouya (cf. Chiche 1990).

The Urban Peripheries and the Fringes of Large Irrigation Sectors: New Systems of Intensive Livestock Production Dominate and Slowly Win the Rural Markets

At the periphery of large and small cities, in the large irrigational sectors, and in the vegetable belts, the keeping of dairy cows, the fattening of bulls and rams for butchering, and battery poultry are practiced. The animals are sold on *souks* or directly to *chevillards*, some of whom are engaged at all levels of the commodity chain. Two sub-systems can be distinguished, namely the mono-activity of grand stables of more than 15 animals, and the creation of small fattening enterprises by small or grand farmers. In any event, livestock production, based essentially on feed from agriculture and industry and requiring veterinary care, profoundly depends on the market. The good condition of these herds witnesses to this. These systems also entail various effects, such as the augmentation of milk consumption in towns, increased movements of products and circulation of money, the dynamization of cooling and fodder industries, the progressive use of veterinary products, and finally the creation of jobs for specialized workers. The majority of pastoralists, after two decades of more or less successful experiments, have opted for cost minimization to such a degree, where it permits them a stock management compatible with an acceptable place on the market. I will analyze below how this branch of livestock is excessively exposed to competition from similar, but more competitive products.

Market Integration and Livestock Production

In the course of the 20th century, the place of animal husbandry has, as it was shown, evolved in the sense of a strong integration into the market economy. It has passed from the status of direct subsistence and unequal social and economic exchange to that of capital, which can be monetarized and which thus assures the rurals' relationship with the market, because it can be exchanged for a wide range of commercial products which the livestock producers need to survive.[20]

In the past, it was only in agricultural areas that everybody could possess their own small herd. Everywhere else, there was a sharp delineation between the *chefs de tente* (herd owners) and their shepherds. The latter were reimbursed by a part of the herd growth, ranging from 10% to a third, and occasionally with wool or fur. Live animals were marketed, as was raw or refined wool, which was delivered to craftsmen or merchants. Until very recently, the offer of animals was high at the time of crisis, whereas demand reached its peak at times of reconstitution of herds after droughts.

The herds were a living capital which had to be safeguarded. Herd capital witnessed substantial fluctuations according to factors which were virtually impossible to control, with which livestock producers had to arrange themselves on the basis of a permanent vigilance towards natural or social dynamics in the greatest radius possible. The knowledge of bio-climatic and social limitations of the most currently frequented pastures thus had to accompany the prospecting for resources and the systematic forging of exchange ties with other regions. For the majority, these relationships fluctuated in parallel to regional climatic and political variations.

Certain livestock producers, however, could establish a more secure livelihood, with revenues escaping the risks inherent to pastoralism. This essentially concerned the families engaged in religious privileges or exercising commercial activities. The first received a tribute or a particular pasture right, because of their believed competence for religious or magical intervention in the face of sickness or calamity, or of arbitration between conflicting parties. It must be emphasized that these goods were not directly redistributed in times of scarcity. Until the last third of the past century, this type of relationship dominated by the *zaouias* has persevered in many regions, for example in the Abda, in eastern Morocco, or the High Atlas. The merchants, in contrast, were the main suppliers of means of subsistence and political protection in exchange for a work as shepherd or laborer, reimbursed by a tithe or a third of produce.

The access to pastures on the basis of reciprocity, stressed in all retrospective studies, seems to have existed in all epochs. Nevertheless, it seems to have been pre-eminently unilateral, according to factors such as the social cohesion of the concerned communities, their respective herd size, and pasture quality.

In the course of the last third of the 20th century, animals reserved for slaughter and fresh milk have become the principal merchandise for livestock producers. That permits them to dispose of regular monetary revenues, but it also obliges them to undertake regular expenditures in order to secure the alimentation needed to reach their production goals. Under these conditions the pastoralists are no longer the leading actors on the market. In the animal husbandry sector and its entire economy, the key positions are now occupied by actors from the trade and service sector, namely *maquignons, chevillards*, manufacturers and merchants of animal fodder, shippers, importers, and retail traders. Today all livestock producers are obliged to go via the market for the raising and commercialization of their livestock, for transport, fodder, troughs and medical check-ups. The most productive branch of livestock production, namely the intensive one, is all the more sensitive to competition through similar, more up-to-par products, because production costs are particularly high due to the use of alimentary products and industrially produced pharmaceuticals. This is as valid for milk, which suffers from the competition of diverse dairy products on the market, as it is for meat, since the expansion of battery-produced poultry and the increased sale of fish in the *souks*. The Moroccan red meat market is, in effect, protected against deep-frozen meat imports. The respective regulations, prone to revision, are founded on several justifications, but since the beginning of the century they require much action led by pressure groups resolutely devoted to the defense against the competition of foreign products.

The market orientation of all Moroccan livestock production systems has encompassed profound modifications in raising techniques, in the various strategies articulating animal husbandry with other activities, and finally in the production and consumption budgets of households. They can be classified in three major types, according to the economic status of the herds and the means of their maintenance:

First, the grand intensive animal husbandry can be qualified as 'the animal husbandry of the rich for the rich'. It addresses a class of mainly urban clients with elevated purchasing power, whose consumption patterns follow models and qualitative criteria established in the European meat sector. Its maintenance is very costly, while the gains vary depending on price fluctuations on the relevant markets. Thus, intensive animal husbandry often requires substantial financial means which may come from different sources, own or exterior, such as revenues from non-pastoral activities, borrowings, or bank credits. Its place in the enterprise economy is quite regularly that of the dependent branch.

Second, the small agricultural workers as well as the middle or grand agricultural entrepreneurs of the cereal regions and the valleys of the northern front of the High Atlas, conduct their animal husbandry with semi-extensive techniques. Both valorize the spontaneous local vegetation and a part of the products from agriculture, using familial workforce. The domestic livestock production in the oases, naturally intensive, is conducted according to the same principle of returning to the market as little as possible. The consequence is a high correlation between a person's general economic condition and the size of his herd. The herds, independently of their size, serve a double economic objective. On the one hand, they are a self-reproducing capital, which works as a treasury fund. On the other hand, the herd is a saving, which can be rapidly monetarized and progressively reconstituted at the pace of financial means. Additionally, the level of gains from the sale of livestock can be controlled by raising several flocks of stock on different parcels, each one with a distinct degree of intensity. One of these flocks, for example, may be destined for commercialization, consisting of selected animals which may be stall-fed. The other flocks may be reserved for needy rural consumers who are content with raw and simple food. These flocks are often composed of weak animals, raised at the minimum costs. In summary, it can be said that this part of the production is conducted by the 'poor for the poor'.

Third, the large pastoral animal husbandries are currently in a transitory phase marked by strong structural transformations. In some regions, herd owners are already settled since 50 years, in others since 20 and in some since five or less years. Many herds are relocated by truck between increasingly distant pasture grounds. But nevertheless, the same fundamental transformations have affected these new sub-systems. A short while ago, these livestock producers tried to guarantee their security by a flexible management of their needs and resources. Their capacity to withstand precarity and risks, and then to undertake a relaunch, was based on continuous monitoring of and adaption to natural and societal dynamics. Their preoccupation was twofold, the well-being of the animals on the one hand, and the subsistence of their social unity on the other. Thus they integrated deaths, losses, offsets, and departures as acceptable in a process of adaptation of the human society in harmony with the dynamics of nature, which is falsely perceived as fatalism or resignation by some. Today the technical and economic management is more concerned about the commercial value of livestock, the relation between costs and gains in terms of profits and losses.

In all of Morocco, pastoralists have lost their power as independent actors. In the traditional systems, even if their control of the physical and social conditions was marked by incertitude, they had a relatively strong hold on the networks and the chains of relationships in which they were immersed as individuals, and none escaped from the obligation to apply evasive tactics when catastrophe struck (cf. Chiche 1998). Presently, economic profit is most often realized in sectors not directly related to livestock production. Pastoralists' everyday life and their social relations thus depend on decisions taken in towns and cities, according to logics which are foreign to the needs of families' survival.

Conclusion

We have seen that in the new economic and social context it is now possible for livestock producers to reduce the vulnerability of their activity vis-à-vis climatic risks. The territorial localization of pastoralism has been modified and new technologies have overtaken the forms of traditional animal husbandry. Henceforward, livestock production remains at the mercy of international and national markets and it is intricately intertwined with revenue systems integrating diverse activities. Nevertheless, pastoral activities still occupy a key position in the consumption and investment budgets of a large part of the rural Moroccan population. Small animal husbandry associated with other activities guarantees the survival and the security of families with weak revenues. For these rural poor, urbanization and agricultural intensification have augmented the mobility of people, products and herds. Moroccan livelihoods are almost always secured by individual mobility under the constraints of vulnerability. But vulnerability has changed its origin. Formerly, it arose from uncertain production and the risk of famines, epidemics, and local political battles. Today, vulnerability is rooted in the need to cover monetary needs indispensable to daily life in a market-dominated society. In this line of thought, it has to be stressed here that in addition to agricultural intensification, the determinant factors for vulnerability are population growth under the constraints of poverty, the increasing occupation of pastoral space, and the concentration of land ownership. The greatest risks for the future of livestock production in Morocco are those which arise from the actors' inability to control the market in line with the recent international commercial accords, into which Morocco is increasingly integrating. Morocco is at the brink of opening national markets to imports without definite delineation of the kind and volumes of imported products. In the light of the current world market situation and the production policies of the world market's main actors – especially Europe, the USA, Canada, Australia and New Zealand – it can be predicted that Moroccan products will face difficulties in competing not only on the global market, but also on its own.

Notes

(1) The following assertions are based on the results of the censuses of 1931, 1960 and 1994. Lack of clarity in the distinction between sedentary and nomadic populations prevented me from using the results of other censuses such as those of 1936 and 1982. The dynamics of the past eleven years, from 1994 to 2005, have been documented by numerous personal observations from which I will draw. Although the data base is inhomogeneous (comparisons between census results have to deal with different units of analysis: in 1931, the number of tents was counted, while in 1960 and 1994 individuals were considered as nomads, basing essentially on their self-declaration), cross-checking with other data about events and situations related in contemporary documents and in oral reports shows that differences in census design do not constitute a major source of interpretative error.

(2) The *guich* were groups of military reservists, recruited from displaced persons. They could also be constituted of amassed population in places of strategic importance for the *makhzen*, the dynasty in power.

(3) Agricultural producers working on a shareholding basis, receiving a fifth of the production.

(4) Sometimes, the settlement of one family can even encompass two or three sites.

(5) This includes the southern slopes of the High Atlas.

(6) For an analysis and discussion of the legal regulations concerning land use in Morocco, see Chiche (1996b). Compare also Bouderbala et al. (1992).

(7) For an in-depth analysis of the complex web of – often diverging and conflictual – interests among pastoral land users in the Ouarzazate region, see Chiche (2004a)

(8) Information about camels is too sporadic and imprecise to be taken into consideration.

(9) For an evaluation of the effects state support programmes have had on sheep and goat production in Morocco, see Chiche (2001).

(10) Supplementary feeding, while important for cattle and draft animals and seldom adopted for goats, especially concerns sheep. For these, the amount of fodder, since the beginning of its utilization, does not exceed 3.6 kg of barley and bran per ten nursing or pregnant ewes, and one kilogram for a ram in the mating season.

(11) All these data are based on Chiche (1999) and HCP (2001), completed by personal observations.

(12) Includes both industrial poultry and poultry originating from farms. The annual consumption of eggs amounts to 90 per person in towns and 34 in rural areas. This is very low when put in a context where egg-consumption could be a palliative to the paucity of meat and dairy products.

(13) Annual per capita meat consumption in rural areas: poultry 4.58 kg, beef 3.9 kg, mutton 3.65 kg and goat meat 0.90 kg.

(14) In equipping these centers systematically, the state aimed at moderating the massive emigration towards large cities.

(15) Office Régional de Mise en Valeur Agricole du Tafilalt.

(16) Regions of Zemmour, Zaer, Chaouia, Doukkala, and Abda.

(17) Except a tiny minority.

(18) Most frequently at intensive agricultural enterprises in the country's north and south.

(19) The ration remains limited to 3.6 kg of a mixture of barley and bran for ten nursing or pregnant ewes, or for mature rams before their sale.

(20) For a review of market integration patterns in Moroccan sheep and goat production systems, see Chiche (2004b).

–4–

Neoliberalism, Environmentalism, and Agricultural Restructuring in Dryland Morocco

Diana K. Davis[1]

On 22 July 2004 the United States ratified the bilateral free trade agreement (FTA) concluded between Morocco and the U.S. earlier that spring (Anon 2004; USTR 2004). This agreement is only the second FTA in the Arab world, following Jordan's agreement with the U.S. which was signed in 2000. The fanfare of press releases surrounding the conclusion of the FTA does not include, however, any mention of the last two decades of neoliberal restructuring in Morocco, including a decade of structural adjustment that paved the way for this agreement.[2] Nor has there been any discussion of the fact that, despite this neoliberal restructuring, Morocco's "growth trend rate [has] continued to slacken, entailing a stagnation in per capita incomes over the last decade, as well as increased poverty and urban unemployment, in sharp contrast with the performance of most other middle income countries over the same period" (World Bank 2001). With at least 40% of Moroccans living in poverty or 'economically vulnerable' and urban unemployment at 20%, the World Bank has recently concluded that "if the economy does not accelerate and new jobs do not materialize during the coming decade, poverty and exclusion are likely to spark unmanageable social tensions" (World Bank 2005a, i).

Such negative effects of neoliberalism and structural adjustment on the social, and often the economic, sectors of society have been documented for many years (Brohman 1996; Peet 1999; Rapley 2004). Because neoliberalism in general, and structural adjustment in particular, tend to enforce a reduction in government involvement at many levels and a reliance on 'free market' mechanisms for the functioning of the economy and society, poor populations are usually hard hit (Brohman 1996, 132–72). State subsidies for staple foods such as flour and cooking oil are often reduced or eliminated, spending on health care and education is rolled back, and lowered tariffs encourage cheap imports that frequently undercut local production resulting in job losses. In addition to these measures of fiscal discipline, trade and financial liberalization, and tax reform, neoliberal restructuring also entails privatization, deregulation and the procurement of property rights of different kinds (Peet 1999, 48–53). Many of these changes necessitate legal and institutional reform which is frequently facilitated by the main multilateral actors such as the World Bank and the International Monetary Fund (IMF) (Swyngedouw 2005).

Neoliberalism, then, is as much a political project as it is an economic and developmental project.

Despite the large and growing body of literature documenting the various effects of neoliberalism on people and the environment, however, scholars have only recently begun to examine, explicitly, the complex relationship between neoliberalism and environmentalism, or how these relations are impacting national development in important ways. It has recently been argued, though, that neoliberalism is necessarily an environmental project, in addition to being an economic, political, and ideological project (McCarthy and Prudham 2004). One of the ways that neoliberalism operates as an environmental project is through its use of environmental narratives, stories that are often derived from the colonial period.

This essay explores neoliberalism, environmentalism (in the form of questionable environmental narratives), and agricultural restructuring in the arid regions of rural Morocco. Specifically, it interrogates the appropriation of a declensionist colonial environmental narrative to help justify the neoliberal goals of land privatization and the intensification of agricultural production in the name of efficiency and environmental protection.[3] In doing so it describes many of the effects of neoliberal reforms in Morocco and analyzes the associated restructuring of dryland agricultural development. Contrary to the commonly anticipated outcome of the 'retreat of government' under neoliberalism, in Morocco some neoliberal reforms have instead consolidated and expanded certain aspects of state power. Relatively little geographical research has analyzed the effects of either environmental narratives or neoliberalism in the Middle East and North Africa. No research to date has examined the complex interaction between neoliberalism and environmental narratives and its impact in Morocco.

The essay opens with an historical overview of neoliberalism and structural adjustment in Morocco. A discussion of (neo)liberal environmental narratives, in the form of questionable degradation narratives, and their colonial precedents is subsequently provided. This is accompanied by an analysis of how these narratives have been utilized to enact neoliberal changes in both policies and legislation, with a specific focus on rural agricultural development. Next, research from southern Morocco illustrates how the narratives have been employed in a predominantly pastoral region, in the name of environmental protection, to further neoliberal restructuring, and to control local pastoralist populations long deemed problematic by the Moroccan government.[4] Land degradation in this region is ubiquitously blamed on livestock overgrazing, an old colonial story which drives significant policy decisions. Despite this rhetoric, however, the documented causes of significant environmental degradation in the region, as in much of Morocco, are the plowing of marginal lands and over-irrigation. Nevertheless, the degradation narrative is being deployed to justify the expansion of state control over collective grazing lands, to encourage privatization, and to intensify production. The paper concludes with a discussion of how neoliberalism and environmental narratives in

Morocco, and their complex interactions, are affecting environment and development in important, and sometimes unexpected, ways.

Neoliberalism and Structural Adjustment in Morocco

As a former protectorate, under French administration from 1912 to 1956, Morocco gained independence with several structural features which have impacted its development over the subsequent decades. The export-oriented agricultural sector, and the way it is integrated into the global economy, are among the most important of these legacies. Another is the way power is wielded through the client system of rural and urban notables, effectively preventing meaningful land reform (Payne 1986). Like the colonial administration before him, the leader of independent Morocco for four decades (1961–1999), King Hassan II, chose to use the best and most developed agricultural land for the production of export crops such as citrus and vegetables. Some of the earliest loans from the World Bank were provided in 1964 to expand irrigated agriculture according to plans originally made during the colonial period (Swearingen 1987). Investment and development of the 'traditional,' non-irrigated sector stagnated. The production of staple cereals, in particular, received little investment and ever larger cereal imports were necessary to provide enough bread for the growing population. These and other imports, poor cereal harvests, and other expenditures by the new government led to chronic deficits and economic crises early in the post-colonial period. As early as 1964 Morocco was receiving funds and liberal economic advice from the IMF (Pennell 2000).

In the early to mid-1970s the Moroccan government increased public spending by several hundred percent. This was stimulated, in part, by the tripling of international phosphate prices which resulted in large state profits from Moroccan phosphate production (Denoeux and Maghraoui 1998; Pennell 2000). In 1975 Morocco claimed the Western Sahara. By 1976, Moroccan coffers were being drained at an estimated one million dollars a day at the same time the bottom dropped out of the international phosphate market (Coupe 1997). The economic crisis continued to deepen and the gap between the rich and the poor continued to widen. In 1978 when the population of Morocco was estimated to be approximately 19 million people, "68 individuals or families controlled 55 percent of private industrial capital, of which one-third was in the hands of only ten groups" (Daoud 1990, 86). The government's economic condition continued to deteriorate and total state debt had risen from approximately 20% of GDP in 1975 to 84% of GDP in 1983 (Coupe 1997, 151; Denoeux and Maghraoui 1998, 57). Foreign exchange reserves were nearly exhausted and for all intents and purposes 'the country was bankrupt' (Denoeux and Maghraoui 1998, 58). By fall 1983 Morocco had entered a structural adjustment program with the IMF which provided a loan of 200 million

USD special drawing rights (ibid.). Shortly after this, the World Bank granted a 600 million dollar package of sectoral adjustment loans to Morocco and many of the country's outstanding debts were rescheduled (ibid.).

This marks the beginning of a decade of structural adjustment in Morocco under IMF supervision. Since 1983, Morocco has entered nine IMF adjustment agreements (Pfeifer 1999). Morocco has since enacted a reduction in price controls and state subsidies, and the liberalization of imports by lowering tariffs and reducing restrictions on import licenses. Funding to the social sectors was cut. Morocco joined GATT in 1987. Exports were also heavily promoted and export duties and licenses were abolished (Denoeux and Maghraoui 1998). The currency was devalued, foreign investment was encouraged and a new investment code to facilitate foreign investment was adopted in 1988. Finally, Morocco embarked on an ambitious program of privatization. Since the official end of structural adjustment in 1992, neoliberal reforms have continued with advice from the IMF and a succession of loans from the World Bank and the European Union (EU). Other institutions including various European governments, USAID and the UNDP have also been important actors in promoting continued neoliberal reforms in Morocco to the present.

The results of restructuring have been mixed at best and very uneven, both economically and socially, over the two decades since reform began. Within the first two years, by 1985, total debt increased from 84% of GDP to approximately 135% of GDP (Coupe 1997). Total debt has since slowly decreased to approximately 40% of GDP in 2003 (World Bank 2005c), but tends to fluctuate widely. Total government debt grew strongly to around 22 billion dollars in the early 1990s and has since stabilized at just about 19 billion dollars for the period 2001–2004 (World Bank 2003a; World Bank 2005b). Debt service payments also peaked in the early 1990s but after falling in the late 1990s have risen steadily for the last several years, nearly regaining their highs of the early 1990s (World Bank 2005c). Although Morocco's balance of payments did improve for a period and the trade deficit has been reduced since restructuring began, both poverty and economic inequality have grown. Much of the improvement in Morocco's finances in the 1990's, though, was due to Saudi Arabia canceling three billion dollars of its debt in 1991 as a reward for help during the first Gulf War (EIU 1999, 10, 38). Such receipts are unsustainable as are those from selling state industries during the phase of privatization. Since 2000, the budget deficit has increased dramatically with current estimates placing it at between 5–6% of GDP (EIU 2003, 26; EIU 2005; World Bank 2005a).

Since the end of the period of structural adjustment in 1992, important living standards have been declining in Morocco, largely as a result of continued neoliberal reforms, despite modest gains in some areas (EIU 1999, 24). Improvements have been small and incremental, but steady, in life expectancy (which has risen from 59 in 1985 to 69 in 2004), GDP per capita (which has increased from $560 in

1985 to $1,520 in 2004 in USD), and infant mortality (which has declined from 90 deaths per 1,000 in 1985 to 36 deaths per 1,000 in 2004) (World Bank 1987; World Bank 2005c). On the ground in Morocco, though, these statistical averages mask the fact, for example, that some Moroccan areas suffer dramatic infant mortality rates (World Bank 2004, 33). Other development indicators, however, reveal worsening trends and inequality has increased overall. Poverty levels, for example, have recently regained their pre-adjustment levels of around 20% after declining in the early 1990s (EIU 1999; EIU 2005; World Bank 1997; World Bank 2005b). In rural areas, poverty is estimated to be 27%, and in some parts of the country it reaches 37% (EIU 2005, 33). Extreme poverty in rural areas tripled during the 1990s and 45% of the entire Moroccan population is now classified as vulnerable to poverty (World Bank 2003b). Seventy two percent of the poor live in rural areas (Anon 2000).

After a decade of reform, literacy rates, reported by the United Nations, actually fell from a high of about 50% in 1990 to a low of about 41% in 1993, a significant decrease that only regained earlier highs in 2001 when literacy rates once again reached 50%, where they remain (UNDP 2005). More disturbing, and in parts of the country perhaps more indicative of declines, a local survey found that literacy rates fell 10% from an estimated high of 65% in 1982 to 55% in 1994 (EIU 1999, 12). Several large provinces in the southwest and northeast of the country currently have illiteracy rates of well over 70% (World Bank 2004, 33). Unemployment has also increased and is estimated to be about 19%, although estimates vary widely from around 12% to much greater than 20% (Cohen 2003; EIU 2005; Pfeifer 1999). Problems of underemployment, not reflected in official statistics, are common in rural areas. For a majority of Moroccans, then, especially the rural poor, life has not improved much, if at all, under neoliberalism.[5] For many, current poverty, unemployment, and illiteracy levels are all too similar to those in the early 1980s before reform began.

Although it has been touted as an 'IMF success story', Morocco's performance in a wide array of social and economic areas lags behind that of the other 'lower-middle income' countries. It also ranks significantly below other Arab states, including its nearest neighbors Algeria and Tunisia, in these areas. Morocco is ranked far below these two Maghreb countries in the human development index (HDI) (UNDP 2005).

Morocco has the lowest life expectancy, the lowest literacy rates, and the lowest GDP/capita of the three Maghreb countries and it also has the highest rate of infant mortality (cf. Table 4-1). Furthermore, Morocco lags behind Algeria, Tunisia, Libya and Egypt in terms of access to adequate health care, safe water and sanitation (EIU 1999, 13). Estimates are that 62% of rural Moroccans do not have access to potable water and 68% do not have electricity (Anon 2000). An estimated 10% of Moroccans live in shanty towns, the highest proportion of all the Maghreb countries (Denoeux and Maghraoui 1998).

Structural adjustment and liberalization of the economy, then, far from delivering 'development' in Morocco, have not succeeded in several key areas. This project has not been a resounding success at the economic level and it has failed at the social level for large sections of society, thus "large segments of the population remain socially and economically marginalized" (World Bank 2005a, i). Some Moroccans, though, have benefited greatly from these reforms. As a result of the very uneven distribution of the few benefits of neoliberal reforms, economic inequality in Morocco has worsened over the last two decades. The main beneficiaries have been the elites long connected with the government. In fact it is striking "how many government elites have converted from the gospel of dirigisme to liberalism" (Dillman 2001, 202). Importantly, "selective privatization has reconfigured and consolidated state control over strategic resources" (ibid., 207). One of the most influential companies with direct ties to the ruling family, for example, Omnium Nord Africain, amongst the largest enterprises in Africa, has purchased many of the Moroccan companies privatized (Coupe 1997; Dillman 2001). This kind of 'privatization' may have very different effects than those intended.

Moroccan elites have long benefited from government enterprise. A wave of new legislation has helped, in many ways, to facilitate the neoliberal reforms and their elite benefits from the 1990s to the present. The privatization of state firms, for example, would not have occurred without the privatization law of 1990 which mandates the privatization of 40% of state companies (Day 2004; World Bank 1997, 48, 81). Many, if not most, of these laws have been pushed through by, and with the assistance of, international financial institutions, especially the World Bank. Many of these laws governing wages, labor, customs taxes, and so forth have received attention and critical discussion in the media and the academy (Catusse 2000). Others such as those affecting agricultural development, water, the environment and land tenure, however, have not received much attention. They are all, however, affecting development in Morocco in important ways.

(Neo)Liberal Environmental Narratives and Agricultural Restructuring

In the rural, agricultural areas of Morocco, where 70% of the poor live, significant changes are being set in place, which are affecting millions of Moroccans, based on an old story of environmental ruin and 'native' improvidence. As told by the World Bank today, the primary cause of poverty in Morocco is "the depleted resource base, degraded and misused for generations" (World Bank 2003b, 23). The assumption that Moroccans have abused their environment for generations to the point that it is degraded and in need of 'improvement' is ubiquitous in Morocco.

Table 4-1: Development Indicators for the Maghreb (2003–2004)

	HDI	Literacy	Life Expect.	Infant Mort.	*GDP/ Capita	**Debt Serv.
Morocco	0.63	51%	70	36/1000	$4,004	9.8
Algeria	0.72	70%	71	35/1000	$6,107	6.5
Tunisia	0.75	72%	74	19/1000	$7,161	6.4
Arab States	0.68	64%	67	48/1000	$5,685	2.5
Mid-Income	0.77	90%	70	29/1000	$6,104	6.4

Source: UNDP, 2005. *GDP/Capita (PPP USD); **Debt service as percent of GDP.

It undergirds many analyses and plans of different kinds from agriculture to forestry to anti-desertification campaigns (Maroc 2001b). It also directly informs policy changes and legal reform, which are increasingly important as neoliberal restructuring continues in the country. One of the main culprits in this story of environmental ruin is the Moroccan herder whose 'over-abundant livestock' are accused of causing overgrazing, 'erosion and soil-degradation' (World Bank 1997, 46). This particular narrative blaming pastoralists for environmental degradation is currently being used to transform agricultural development, including land tenure and use, in the non-irrigated, 'rainfed' areas of Morocco. It is important to recognize, however, that this narrative is nearly two centuries old and has been used for a very long time to appropriate land and resources from the less powerful segments of North African populations as well as to control their actions. The relations between environmental narratives and neoliberalism in Morocco, then, have very deep and complicated roots.

The origins of this narrative lie in the history of French colonial Algeria. When the French occupied Algeria in 1830, the story rapidly took shape that during the Roman period, nearly two thousand years earlier, North Africa had been covered with thick forests and was so fertile that it had become the granary of the Roman empire. The second half of this story blamed the Arab nomad invaders of the eleventh century and their rapacious herds of livestock for ruining, deforesting and desertifying the land. The arid environment they found in Algeria, which had relatively sparse vegetation, the French erroneously assumed was the result of overgrazing and abuse continued unabated during the subsequent centuries. This environmental narrative was used by the colonial administration to help appropriate nearly all forested land in Algeria in the name of environmental protection, and to appropriate much of the best agricultural and grazing lands during the colonial period (Davis, forthcoming). Furthermore, this narrative was invoked to justify the sedentarization of most of Algeria's nomads and to control their movements (Davis 2004). This had the additional effect of creating a large labor pool to supply the need for workers in the expanding capitalist agricultural production system of nineteenth century Algeria. The narrative was also used to change and rewrite numerous laws and policies over the course of the colonial period. In the process, the traditional uses of the forest and other lands by the Algerians were systematically criminalized and the majority of the indigenous population was marginalized and impoverished. The same environmental narrative was carried to Tunisia in 1881 and to Morocco in 1912, with much the same effect (Davis 2005b).

One of the most important impacts of French colonial administration in North Africa was the transformation of subsistence production by the local population to commodity production primarily by European colonists and companies. Forests which had provided pastures, foods, medicinal plants, places to live, and agricultural plots were demarcated, rationalized and improved to provide timber, cork and other forest products for market. Pastures which had supported millions of sheep and goats and allowed local herders to subsist on milk products and to trade or sell wool and hides for grains and other necessities were turned into sites of modern, intensive sheep raising to provide tender lamb and fine wool for the French consumer. Agricultural production, which had been largely subsistence oriented, was modernized to produce for the market, both domestic and foreign, and it required land. The transformation to this type of commodity production for the growing global economy of the nineteenth and early twentieth centuries occurred during the height of classical liberalism.

It is widely recognized that one of the major tools used during liberal economic expansion was the 'enclosing of the commons' which amounted to appropriating/privatizing common land, and restructuring property relations and property distribution (Thompson 1993; Williams 1973). This could not be accomplished in North Africa, however, in the same way it was in various European countries, most notoriously, as it had been accomplished in England. The French were faced with the significant problem of first appropriating land, and not simply of reconfiguring pre-existing property relationships to procure land. Using liberal ideological arguments such as the notion that 'unimproved' land was value-less (waste land) and that land should only be owned by those who make it productive, the French added the story of environmental ruin by the 'natives' to fashion very powerful justifications for land and resource appropriation. First in Algeria, and later in Tunisia and Morocco, this liberal argument favoring 'productive' private property was combined with the narrative of the ruin of a previously fertile land by Arab nomads and their herds to justify many laws and policies which allowed the expropriation of vast amounts of forest, agricultural and pastoral lands during the colonial period. The two stories thus reinforced each other in an interesting example of liberal environmentalism which resulted in far reaching changes in French North Africa.

This declensionist colonial environmental narrative has been embraced by the post-colonial governments in North Africa to varying degrees and for different purposes. Over the past few decades it has also been eagerly appropriated by institutional actors promoting neoliberalism in Morocco. USAID, for example, has been utilizing the narrative to encourage reforms in rangeland management for 30 years. Invoking Garret Hardin's liberal 'tragedy of the commons' thesis, which claims that all common land will necessarily be over exploited and thus should be privatized, USAID has strongly recommended that in Morocco "the collective pastures must first be enclosed" (Gow et al. 1985b, 1).[6] USAID claims that enclosing the collective pastures is necessary due to severe degradation, although it admits that "there are *no reliable data* on the degree of present degradation" (Gow et al. 1985a, 11; Gow et al. 1985b, emphasis mine). Despite this disturbing lack of data demonstrating overgrazing and land degradation, USAID has been advising Morocco to privatize rangeland since the 1960s.

Importantly, USAID also 'assisted' Morocco, along with the World Bank, in writing the agricultural law, the Agricultural Investment Code (Code des Investissements Agricoles), in 1969 (Gow et al. 1985b, 2; Swearingen 1987, 161). This complex law covers nearly every aspect of Moroccan agriculture with a particular focus on the high-value irrigated areas. The code creates irrigation perimeters in the irrigated areas which it governs with very specific rules of use (Delannoy 1969; El Khyari 1987; Swearingen 1987). USAID contributed to the sections treating rangeland management and livestock development which reflect USAID's predominant goals of privatizing rangeland and intensifying production for market. One of the most important aims of the 1969 code was stated baldly as 'the progressive suppression of [collective lands]' (Delannoy 1969, 153). Within the irrigation perimeters, collective lands were privatized (Delannoy 1969, 153; Swearingen 1987, 164).

This 1969 code also defined perimeters of pastoral improvement (périmètres d'amélioration pastorale) which "provided the legal basis for the creation of range improvement perimeters on communal lands, thereby ceding control over the management and development of these perimeters to the GOM [government of Morocco]" (Gow et al. 1985b, 2). This theoretically nationalized sections of Morocco's rangelands said to be degraded, in order to "fight against the degradation of pastures" (Delannoy 1969, 155). It was not until the 1980s and later, however, that results of this agricultural code began to be seen in rangeland areas. By 1982, for example, only about 180,000 ha had been delimited as pastoral perimeters, a very small portion of Morocco's estimated 21 million ha of rangeland (El Khyari 1987, 387; Maroc 2001b, 21). The majority of collective lands still extant at the time of independence retained the status of collective lands into the 1990s.

Of Morocco's approximately 71 million ha of land, roughly 11 million ha, or 15%, are considered collective lands (Maroc 2001a).[7] About 14% of these collective lands, 1,544,656 ha, are arable with most of the rest constituting rangelands (Maroc 2001b). The 9.5 million ha of collective rangelands constitute nearly half of the country's estimated 21 million ha of 'manageable' rangeland (ibid.). Arable land in Morocco covers about 8.5 million ha and only about one million ha are irrigated. Nearly all the unirrigated land is planted in cereals in varying amounts depending on rainfall. A substantial amount of the arable land, roughly 18–19% is collective land (Maroc 2001a; Maroc 2001b). It is impossible to calculate how many Moroccans benefit from the collective lands since the government does not keep statistics of this kind. A conservative estimate of these numbers would be in the millions, perhaps as high as 10% or more of the current population of 30.6 million people (World Bank 2005b). In addition to supporting agricultural and livestock production, the collective lands also provide basic necessities such as gathered foods and medicinal products, as well as materials for domestic building and fuel needs. Moreover, they "play a central role in the maintenance of the peasantry in the countryside and in their resistance to dispossession" (Bouderbala 1997, 160).

With 18–19% of arable land and 45% of rangeland under collective ownership, a lot is at stake for the millions of beneficiaries of these collective lands if collective property rights are eroded. Over the last decade, under the guidance of the World Bank, and with the agreement of the Moroccan administration, this is precisely what has been taking place. The World Bank, it could be argued, has been disproportionately influential in Morocco since the 1960s. It has orchestrated much of the restructuring of Moroccan agriculture over the last 40 years (Swearingen 1987; World Bank 1994b; World Bank 2001).[8] In the mid 1990s the Bank facilitated a major revision of Morocco's agricultural code, effectively rewriting much of it as a more liberal, interventionist and widely applied law. The new agricultural code addresses Morocco's non-irrigated areas and reflects many liberal goals long held by the World Bank for transforming Moroccan agriculture and its rural development. It also clearly invokes the colonial narrative of environmental ruin from 'native' misuse or ignorance. There has been an urgent need to intervene, according to the Bank, because "the quality of the natural resource base is deteriorating: there is severe soil erosion, degradation of the watersheds, depletion of forest cover, overgrazing of the rangelands and depletion of the soils" (World Bank 1994b, 2). Much of the problem, despite scant documentation, is claimed to be caused by serious overgrazing of the 'large herds' kept by pastoralists on the collective lands. According to the Bank, "the maintenance and improvement of the Moroccan rangelands are jeopardized by poor management [and] an unclear definition of user rights on collective land" (World Bank 1994b, 24). Worse, perhaps, in the eyes of the Bank, is that "the lack of a comprehensive land policy has constituted an important constraint to agricultural development and has resulted in an extreme fragmentation of holdings, a lack of land titles, and poor management for certain judicial statutes, in collectively-owned lands in rainfed areas" (World Bank 1994b, 9). The Bank has been trying to address these issues, specifically trying to privatize the collective lands for decades.

The new agricultural code for Morocco's drylands consists primarily of one comprehensive law. This is law number 33–94 governing the perimeters for the development of rainfed land (loi relative aux périmètres de mise in valeur en bour) which was passed in 1995 (Maroc 1995). In various respects it applies to the drylands many of the clauses of the 1969 agricultural code that previously only applied to the irrigated areas (MADR 2000). It breaks down the drylands into three zones: those of agricultural development, pastoral amelioration, and the conservation of soils. Importantly, a fundamental principal of this new law is that exploitation of agricultural land inside the perimeters is obligatory (Maroc 1995). Once a perimeter is declared and defined, the management of land tenure within its borders is declared of public utility and thus comes under control of the State. This means that within these perimeters 'only modes of exploitation conforming to contractual modalities fixed by agricultural legislation are authorized' (Maroc 2001c, 134). This law was accompanied by law number 34–94 of 11 August 1995 governing the limitation of land parceling in both irrigation perimeters and dryland perimeters (Van Buu 1997, 691).

These two laws are exactly the same, in name and purpose, as those advocated in internal World Bank documents and drafted by World Bank staff for the Moroccan government (World Bank 1994a, 58; World Bank 1994b, 9). These World Bank reports, however, also document that a third law was drafted to accompany the two passed in 1995 (33–94 and 34–94). This is a draft law governing the 'privatization of collective rainfed land (projet de loi relatif à la melkisation des terres collectives)' (World Bank 1994a, 58). These three laws are considered by the World Bank to be a necessary part of the process of establishing a land bank and a land market in Morocco. The World Bank states that this legislative framework will encourage the development of the rainfed land through the constitution of economically viable farms and that land consolidation will help reduce farm fragmentation which will increase productivity (World Bank 1994a, 58).

For the government of Morocco, the goal of these laws is to modernize agricultural development in rainfed areas, increase cereal production, and promote exports and private investment (El Asri 1999). The Moroccan government shares with the World Bank most of its neoliberal goals. It also shares with the Bank the narrative of environmental decline at the hands of local Moroccans. As stated in the government's recent five year plan, in order to rationally use the soil and to safeguard agricultural land from degradation, the following measures must be taken, including, the "... institution of legislation permitting the deliverance of private, individual land titles for the beneficiaries of collective lands to encourage their development" (Maroc 1998). Thus in its "strategy of land development to the horizon 2020", the Moroccan state makes clear that it intends to create a land market and to "lift all the legal obstacles to the free disposition of land" (MADR 2000). In its fight against land fragmentation the Moroccan government has encouraged land consolidation and pursued registration on private and collective lands, in addition to surveying half of the national territory (Maroc 2000, 16–7).

Despite the rhetoric of the Moroccan government and the World Bank, recent research has shown that, in Morocco, "farm fragmentation has no impact on agricultural productivity" and that "consolidation and titling programs cannot be expected to automatically increase agricultural productivity" (Benabdellah 1998, iv–v). Moreover, the history of land privatization and attempts by the World Bank to create land markets/banks in other parts of the world have 'often been a disaster' with communities losing their previously communal lands in only a few years (Food First 2002). Likewise, the kind of livestock development being promoted to fight against the presumed degradation of pastures has repeatedly failed over decades of attempts throughout Africa and the Middle East. The largest pastoral project of this type in Morocco, based almost entirely on exclosures and rotations, has recently been judged unsuccessful. The project, indeed, concluded that it is difficult to distinguish between the effects of the climate and the effects of management on vegetation change (Msika 1997, 41). A long history of failed range development projects in Morocco calls into question the narratives of overgrazing and degradation that underpin them (De Haan 1994; El Harizi 1998; USAID et al. 1986). In fact, the most recent government report on the state of the environment in Morocco admits that "the real extent of pasture degradation is difficult to measure ... at the national scale *no quantitative estimation of the state of degradation of rangelands is available*" (Maroc 2001c, 128, emphasis mine). Moreover, the estimates made for Morocco overall (81% of pastures moderately degraded) appear to have been generalized from a study of only 640 hectares of rangeland (Maroc 2001c, 128). The environmental narratives being used to justify neoliberal restructuring of rural agricultural areas, then, appear to be less well founded than the story tellers claim.

Land Degradation in Southern Morocco

The narratives of overgrazing, land degradation and desertification are very pervasive in the southern regions of Morocco. Ouarzazate province is one of these pre-Saharan regions frequently claimed to be suffering "intense degradation caused by overgrazing" (ORMVAO 1993, 1).[9] Collectively owned land constitutes 95% of the land in Ouarzazate province (Faouzi 2002). Ouarzazate is also one of the frontier zones of Morocco, bordering the contentious, ill-defined Algerian border. Many of the pastoral nomads in the region regularly crossed this border in the past, although they can not do so today because the border is heavily militarized. For hundreds of years nomadic tribes from the south threatened and occasionally successfully invaded governments in Morocco (Abun-Nasr 1971; Burke 1976). These tribal groups were also the most difficult for the French to conquer and it took about half of the protectorate period to successfully control and pacify them (Bidwell 1973; Dunn 1977). Thus there is a long legacy of mistrust and surveillance of this and other frontier areas. The declensionist environmental narrative is being used in this region to further neoliberal reforms in land tenure and agricultural development.

Overgrazing is most commonly blamed for the condition of much of the land said to be degraded in Ouarzazate. Government documents report that pasture degradation is intense and severe, that the number of livestock widely exceeds the carrying capacity, and that pastures are used in an irrational manner by the herders (ORMVAO 1990; ORMVAO 1992). These and similar government documents state that due to this degradation grazing has to be rationalized, herds destocked, pastoralists' movements controlled, and production intensified (demobilized). These changes are pursued despite evidence of greater livestock productivity in the mobile pastoralist systems in southern Morocco compared to the sedentary systems (Darfaoui 1995, 64). Numerous projects have been launched in the name of reducing degradation that have severely disrupted pastoralists' livelihoods, often further marginalizing them. Following the national trend outlined above, very little physical evidence is provided in support of these claims (Bendaanoun 1994; El Gharbaoui 1985; Habib 1982; Korachi 1995; PROLUDRA 1997).

What existing ecological studies conducted over the last 25 years do show is that there is scant 'hard' data demonstrating permanent overgrazing or a significant decline in vegetation over time for Morocco's extensive rangelands. Rather, numerous exclosure studies and botanical surveys indicate that vegetation cover changes dramatically over time and that these changes are generally correlated with rainfall more than any other factor. More than one researcher has concluded that "overgrazing seems to have a minor role" in land degradation in the region (Waibel et al. 1993, 19). Furthermore, nomadic pastoralists living in the region report little or no decline in the quality of the pastures on which they rely over the last several decades. The vast majority of those herders interviewed did not believe the land in the region was overgrazed or desertified. Some of the government range personnel also expressed their doubts about desertification being a problem and said that overgrazing was not a significant problem in most areas (Davis 2005a). This is not to say that there is no land degradation in Ouarzazate. It is likely that some limited overgrazing is occurring in localized areas around towns and villages. The majority of significant land degradation, though, is found elsewhere, in the over-irrigation and the plowing of marginal land.

As a result of Morocco's history of building large dams and its irrigation policies, widespread disruption of hydrological systems and traditional water distribution networks has occurred. This has raised the water table in places, causing salinization and water logging, and it has lowered it in others making water more difficult and expensive to obtain. A recent study concluded that nearly half the irrigated area in Morocco, approximately 500,000 ha is affected by salinization (Maroc 2001c, 131). Over extraction of ground water has caused saline intrusion of aquifers in the coastal areas of the southwestern plains (Souss and Tensift) (World Bank 1995, 7). In southwest Morocco, for example, the poor germination and growth of the Argan tree, long blamed on overgrazing by goats, has recently been shown to be due to soil salinization (Michmerhuizen 1999).

In southern Morocco, salinization is a large and growing problem. In the Draa valley (Ouarzazate), salinization of the soil in many areas has increased at an alarming rate. In some of the palmeries along the Draa, salinization affects more that 80% of the agricultural area, with up to 45% of the affected areas having saline soil loads high enough to be very detrimental to agriculture (ORMVAO 1983; Thiault 1994). In some areas nearly 10% of the land has been abandoned altogether (Maroc 2001b, 26). Many of these areas also suffer from over extraction of ground water. In some places the water table has dropped as much as eleven meters in fifteen years (Jardak 1996, 51). Several efforts to create drainage in these areas have failed, resulting in standing, stagnant water that fosters diseases.[10] Part of the reason for the expansion of irrigated agriculture in such an arid region is the government's provision of free water for agriculture (Miller 1996; Jardak 1996; Swearingen 1987). This water subsidy is part of the government's efforts to slow down the exodus of rural Moroccans to urban areas.

The other primary cause of documented land degradation in southern Morocco is the ploughing of marginal soils which disrupts local vegetation and desiccates the soil. A provision in the liberal 1969 Agricultural Investment Code has encouraged this by allowing the appropriation and privatization of collective land through cultivation by members of that tribe (Delannoy 1969). Following a complex set of rules, such land can be claimed as private property (complete with the right to bequeath the property to an heir) if it is equipped with a motorized water pump and intent to cultivate is demonstrated (Jardak 1996). If a well is drilled and not equipped with a motorized pump, ownership is established for at least one year even if the land is not cultivated. Ownership can be claimed indefinitely if the land is cultivated. Importantly, the land can be claimed for six months simply by delimiting a parcel with intent to cultivate (ibid., 152). This, combined with government guaranteed high producer prices for wheat, has encouraged a 161% increase in cereal cultivation, from 1965 to 1994, the majority in marginal areas. In the 'unfavorable' areas of the South, that is the more marginal areas with low rainfall, the area planted in cereals increased 38% from 1991–1996; in the 'Saharan' areas, cereals hectarage increased 24% during the same period (MAMVA 1996, 31).

In Ouarzazate province, the area planted in cereals has increased from 15,000 ha in 1975 to 56,000 ha in 1995, an increase of 273% over 20 years according to the annual reports available from the provincial government. Much of this cereal expansion has taken place on collective lands previously used mainly for grazing. Whether land is planted in cereals or not, in these arid zones, plowing the soil not only destroys the local vegetation, it increases the evaporation of moisture from the soil and effectively increases the risk of drought (Swearingen 1996). As one geographer explains, "[f]armers are destroying the grazing potential of these lands through clearing the natural vegetation, [and] unsuitably 'mining' their fragile soils" (Swearingen and Bencherifa 1996, 28). Cereals planted in this region very often fail (unless irrigated), except in rare years of exceptionally high rainfall, since dryland cereals, especially wheat, generally need at least 400 mm of rain to grow a suitable crop (ibid., 20–1). Ouarzazate province receives an annual average of 250 mm of rain, and half that or less in the southern portions of the province.

This brief case study of land degradation in southern Morocco shows that government agricultural policies have promoted over-irrigation and the expansion of cereal cultivation on marginal lands. Despite the documentation of these practices, attention is being diverted from these sources of serious land degradation by the focus of government and international institutional actors on the questionable crisis narrative of overgrazing and desertification. This environmental narrative is being used to help justify radical changes in land tenure, land use, and agricultural practices, as it was during the colonial period. These changes focus on agricultural intensification, export production and the privatization of collective lands. Moreover, in pastoral areas, the 'ecological need' to destock, reduce and regulate mobility, and to enclose common land reinforces many strategies of a government which has long-standing aims of controlling nomads in frontier areas (Bencherifa 1996).

Discussion and Conclusion

Neoliberalism has been enthusiastically, if selectively, embraced by the Moroccan government and much of the business elite in contrast to the opposition it has encountered elsewhere.[11] Neoliberalism, with its inherent free-market environmentalism, also has reinforced several pre-existing goals of the Moroccan administration in agriculture and rural development. It has done so by legitimizing the government's goals of land privatization and consolidation with an economic narrative of the efficiency of large landholdings under mechanized, intensive production. From this point of view, despite evidence to the contrary (Benabdellah 1998), the fragmented small holdings of the majority of Moroccan farmers, worked with few inputs and on marginal land, are wasteful, inefficient, and in need of improvement and modernization. The relationship between neoliberalism and environmentalism in Morocco, though, is more complex than this. The market environmentalism which pervades much neoliberalism around the globe is accentuated in Morocco by the incorporation of the colonial narrative blaming land degradation on traditional land use practices such as pastoralism and subsistence farming.[12] This narrative originated in colonial Algeria and it facilitated the expropriation of land and resources throughout the nineteenth and early twentieth centuries. It was utilized widely to help appropriate collective lands, a classic instance of enclosing the commons so emblematic of the changing social relations with nature during that period of classical liberalism and the rise of the global economy.

The current use of this declensionist narrative by the Moroccan government and international financial actors has facilitated a contemporary enclosing of the commons, an example of what David Harvey has recently called an act of 'accumulation by dispossession' (Harvey 2003). The utilization of this narrative has also cast Moroccan pastoralists and subsistence farmers as double eco-outlaws. These Moroccans, among the most poverty stricken and powerless in the country, have been portrayed both as ecological outlaws and as economic outlaws despite questionable documentation for either claim.[13] The complex neoliberal environmentalism current in Morocco today blames the pastoralist and small holder for degrading the environment at the same time that it accuses them of inefficient or wasteful modes of agricultural production. This powerful discursive formulation has been used to justify and facilitate the legal revision of the agricultural code and rural development plans for Morocco's drylands as detailed above. In southern Morocco, the declensionist narrative of overgrazing has been used, contrary to evidence that the majority of land degradation is caused by over-irrigation and plowing of marginal soils, to gain control of collective pasture lands and to regulate the resource use and the movements of pastoralists. In parts of Morocco, then, these narratives affect governance in important and complex ways in that they are being used in the service of neoliberalism at the same time that they have expanded state power.

These changes, especially the law governing the development of rainfed land are having a substantial impact. Twenty five rainfed land development perimeters have already been designated covering half a million hectares and affecting 350,000 Moroccans (Maroc 2000; World Bank 2003b). Under this law, plans for the future include 180 projects over three million hectares (El Asri 1999). Immediate plans include 70 projects for agricultural perimeters covering one million hectares and affecting one million people (Maroc 2000, 46). The majority of these projects are likely planned for collective lands as the bulk of rainfed areas are areas of collective property rights. One of the proposed dryland agricultural perimeter projects to be funded by the World Bank, for example, targets six 'priority action zones' of which five are in provinces dominated by collective property rights (World Bank 2003b). Fifteen perimeters for pastoral improvement have already been demarcated covering approximately three million hectares of rangeland (Maroc 2000). Over the next 15 years, the government plans to delimit 40 million hectares of pasture in these perimeters and to enclose 8,300,000 ha in exclosures which prohibit grazing, in addition to creating 814 pastoral cooperatives (MADR 2000). This 40 million ha is nearly double the amount of rangeland considered 'manageable' by the government and would thus include all of the collective pasture lands.

Whether or not these collective rangelands and the 1.5 million ha of agricultural collective lands are ever privatized outright, severe restrictions are enforced by these new dryland perimeters. Land within the perimeters is declared of public utility and comes under the control of the state. The state then dictates certain modes of production which must be carried out, thus criminalizing other modes of production or alternative resource uses. The punishment for infractions of these new rules can include six month's imprisonment and fines that range from 200 dirhams to 5,000 dirhams (Maroc 1995, article 48). These fines represent roughly one week's to nearly five month's work at the average minimum wage (Maroc 2001a). This is high enough to drive many families deep into poverty and to force some to sell their livestock and migrate to the growing shanty towns in the larger cities. For the millions of poor Moroccans who benefit from and rely on the collective lands for medicines, foods, and other basic necessities, then, the imposition of these perimeters and their new rules are having negative effects. The largest pastoral project carried out under these new rules thus far did not significantly improve the state of the pastures; it further exacerbated economic inequality, and it dispossessed some of the poorest herders from their livelihoods and resource base (Msika 1997; Steinmann 2001).

The implementation of the bilateral free trade agreement between Morocco and the United States, ratified in July 2004, will also likely worsen poverty and make life much more difficult for poor Moroccans in the dryland agricultural areas. By mandating the elimination or substantial reduction of export subsidies and price supports for grains in Morocco as a result of the agreement, the United States expects to profitably supply nearly all wheat to Morocco in the near future. It is already recognized that this will likely disrupt rural agricultural production severely, potentially dispossessing large numbers of Moroccans of their livelihoods and increasing poverty in rural areas (Nathan Associates 2003, 37–9). Recent estimates by the World Bank project that poverty in rural areas will increase from 28% to 34% and extreme poverty in rural areas will double from 6% to 12% (World Bank 2004, 62). If alternative employment is not created and social safety nets are not put into place as the trade agreement is implemented, a significant number of the four million Moroccans employed in agriculture in rural areas, as well as their families, will be adversely affected. Given the history of Moroccan governance, these alternatives are unlikely, especially with the reduction of state services under neoliberalism. The trade agreement, then, will likely enhance trends already set in place during the neoliberal restructuring of the past two decades which have exacerbated environmental degradation, increased poverty, and expanded state power in several ways. The only other Arab country which has signed a free trade agreement with the U.S., Jordan, has indeed not experienced the democratic reforms the agreement was expected to deliver. Instead, the free trade agreement in Jordan "has necessitated a steady reversal of the political liberalization welcomed by Jordanians in the late 1980s and early 1990s" (Moore 2003). A similar outcome is likely in Morocco.

By examining the historical, colonial context of some of the dominant environmental narratives in Morocco today and how they relate to neoliberalism, this article has attempted to show that similar narratives have been utilized for nearly 200 years to facilitate (neo)liberal environmental governance in North Africa. Recognizing that these stories of 'native improvidence' and the inefficiency of traditional production systems have been used to dispossess local peoples from their resources and properties, and to transform modes of production for centuries, allows us to examine the present more clearly and more critically. Not only are these environmental narratives being utilized under contemporary neoliberal restructuring to gain control over land and resources, they also facilitate aspects of state authoritarianism and erode democratic reforms. This article suggests, then, contrary to some of our more simplistic notions of neoliberalism, that in Morocco its definition and selective implementation have been quite complex. Neoliberal reforms have encouraged the retrenchment of government in many areas, such as spending on social services, for example, at the same time that it has expanded state power and state control over resources such as the collective pasture lands.

The deployment of false environmental narratives to justify political and economic control over resources and to further interest-laden environmental policies are now well-recognized phenomena, particularly in sub-Saharan Africa (Bassett and Zuéli 2003; Fairhead and Leach 1998; Leach and Mearns 1996; Neumann 1998; Swift 1996; Turner 2003). The Moroccan case detailed here highlights some of the benefits of analyzing the complex relations between such enduring environmental narratives (often packaged as contemporary environmentalism) and neoliberalism. Interrogating these environmental stories, by exposing their frequently poor documentation as well as the motivations for their use, is perhaps one way to begin to transform the neoliberal project and to encourage governance that will benefit both people and the environment.

Notes

(1) An earlier version of this chapter appeared in *The Geographical Journal* 172 (2006), 2, 88–105. Reprinted by permission of Blackwell Publishing.

I am grateful to the many Moroccans who helped me at various stages throughout this research and to the Moroccan government for granting me research permission. The research for this paper would not have been possible without generous funding provided by the US EPA STAR Fellowship program and the Social Science Research Council's IPFP, IDRF, and NMEP fellowships. Charlie Hale and Becky Dolhinow discussed the many complexities and nuances of neoliberalism with me that gave me new insights. I would like to thank Ian Manners, Neil Smith, Susan Slyomovics, Christian Brannstrom, James Housefield, three anonymous reviewers, John Briggs, and Jörg Gertel for their insightful and constructive comments on this paper.

(2) Bush administration sees this new Moroccan FTA as an economic boon to both countries, and also proclaims that it "can show many others throughout the Middle East the power of free trade to support democracy and promote prosperity and build a more tolerant, stable and peaceful world" (Shelby 2004).

(3) 'Declensionist' is used in this article to describe environmental narratives that are essentially stories of degradation, many of which were written during the colonial period and usually blamed indigenous peoples for assumed environmental destruction. The work of Fairhead, Leach and Mearns has been foundational in exposing this kind of narrative and its (mis)uses in sub-Saharan Africa (Fairhead and Leach 1996; Leach and Mearns 1996).

(4) Research for this article was conducted over a period of two non-consecutive years that involved the collection of primary documentary data from Moroccan agricultural ministries and regional offices, interviews, and qualitative research on local knowledge of environmental change in the south of the country. The majority of the research involving local knowledge has been published separately (Davis 2005a).

(5) Some Moroccans have protested the poor conditions in their country, especially low wages. A series of strikes and 'social dialogue' in the late 1990s and early 2000s resulted in the minimum wage being raised more than once (EIU 2005, 33). For some of the social/political ramifications of neoliberalism in Morocco and the sometimes successful protest it has produced, especially in urban areas, see Cohen (2003). This author argues that even the 'modern middle class' has declined since reforms began in the 1980s.

(6) Garret Hardin's work has been enormously influential over the last 30 years although he makes the fundamental error of mistaking open access property for common or collective property. For his views and an excellent critique see (Hardin 1968; Hardin 1974; McCay and Acheson 1987). It should also be noted that Hardin hypothesized that the tragedy of the commons might be avoided as well with strong centralized state controls.

(7) The figure of 71 million hectares of national territory claimed by the Moroccan government includes the territory of the Western Sahara. The area of Morocco without the Western Sahara is approximately 46 million ha (EIU 1999). Statistics on the area of Morocco and land uses in the country vary according to different sources, making comparisons difficult. Best approximations from the most reliable sources are used here.

(8) As of October 2000, the World Bank had made 34 agricultural loans (and dozens of other loans) to Morocco totaling over two billion dollars. Five of the seven loans evaluated over the last 10 years were considered unsatisfactory, and two were ranked highly unsatisfactory (World Bank 2001, Annex A, 23–5).

(9) Ouarzazate province was divided in the 1990's into two different provinces, Ouarzazate province and Zagora province. More recently a further bureaucratic reshuffling created a new, larger administrative region, Sous-Massa-Draa. Ouarzazate province will be used in this paper to indicate the entire region of what is now known as Ouarzazate and Zagora provinces, but not the Sous-Massa region.

(10) Interviews and field research in southern Morocco conducted by the author.

(11) For examples of parts of the neoliberal project that have not been completely implemented in Morocco, see EIU (2003, 25–26).

(12) Market environmentalism is widely understood to be the process of managing the environment based on the value of its salable products, that is, via the commodification of nature (McAfee 1999; McCarthy and Prudham 2004). Morocco's new environmental protection law provides a good example of this kind of environmentalism in Morocco. Promulgated in January 2003, this 'law is no. 11–03 relative to the protection and development of the environment' (*loi no. 11–03 relative à la protection and mise en valeur de l'environnement*).

(13) This formulation draws on the provocative work of Michael Goldman on eco-governmentality and green neoliberalism (Goldman 2001).

Regional Geographies of Pastoral Morocco

Social Change and Sedentarization in the Eastern Moroccan Steppes

Boutayeb Tag

The dissolution of the nomadic way of life in the steppes of the eastern Moroccan plateaus traces back to the last quarter of the past century. Since then, a certain number of actions undertaken by the pastoralists has accompanied this dissolution. These actions are about the transformations of the agro-pastoral system and they reflect an adaptation to an overall development of living conditions dictated by endogenous dynamics and by the impact of globalization. Simultaneously they reflect a strategy of safeguarding social-tribal interests in the framework of the development and the country's political administrative reforms. Some of the most significant transformations concern the ever-increasing integration of agricultural activities into the practice of livestock production, which was formerly the preeminent activity practiced by the locals of the steppes of the eastern Moroccan plateaus. Other aspects of transformation are the diversification and amelioration of livestock and the introduction and proliferation of motorization in this habitat amidst the steppes. It is argued that the overall new degree of market-orientation in pastoralism has led to an increasing social stratification and sedentarization among livestock producers. Mainly the wealthy seem to be prospering and benefit from lucrative new market niches.

Spatial Transformations

The agro-pastoral activity in the region has experienced changes in its structure. These are related to the dissolution of nomadism in this region prior to the 1990s. Recently new spatial transformations emerged in the Oriental. First, Perimeters of Pastoral Amelioration (PAP) were created: For more than three decades the region has witnessed the degradation and denudation of the vegetation cover caused both by the climate and human intervention, like recurrent droughts, the abusive exploitation of the steppe and the burden of pastoral overstocking. In order to limit this destructive influence the state has implanted several zones of pastoral amelioration. This program entails the closing-off of pastures concomitant with the plantation of *Atriplex* on areas comprising each a couple of hectares. Those entitled to use these areas receive a compensatory indemnity. Such spatial management in the region's pastoral field has introduced a new territorial organization and a new management of pasturelands, initiated and controlled by the Waters and Forests Administration (Administration des Eaux et Forêts). The management of these new pastoral spaces is based on a minimum of equipment and on a rotational exploitation. The pastures are closed for three to five years to allow for the regeneration of natural vegetation and/or the plantation of *Atriplex*, then they are reopened to grazing. But what are the PAP's impacts? One consequence is expressed in the changing landscape of the steppes: here one encounters barbed-wire enclosures, demarcations of *Atriplex* and a generally recovered vegetation. Other impacts have a socio-economic character. They will be presented in the subsequent section.

Second, pastoral cooperatives were created: More than forty pastoral cooperatives have been established in the region of the Oriental. If their geographic distribution is less visible on the spatial plan, they nevertheless assume a role that directly intervenes on the pastures which belong to each cooperative. The pasture of herds on these areas is reserved for members. Pastoralists, who do not belong to the cooperative, are excluded, even if they belong to the same tribe. Besides, absent members of the cooperatives lease out their grazing rights to livestock producers who are able to pay.

Third, collective land was appropriated through sedentarization: The nomadic sedentarization in the high plateaus of the Moroccan Oriental is not a recent phenomenon. It is connected to the living conditions of pastoralists. Each time that periods of extreme drought hit the livestock producers and decimated their herds, the most vulnerable ones abandoned nomadism and the steppe. They then settled around the region's *douars* and progressively turned to oasis farming, especially in the *douars* of the Moulouya valley – such as the Oulad Khaoua, Oulad Mellouk, Oulad Jerrar – and in Oued Za as well as in the oases of the southeast. In the early 20th century the colonial administration put in place a system of cantonment, regulating nomadic mobility. This led to the exclusion of some nomads from their pasture. These then had to settle and work as agricultural laborers or shepherds in the service of agro-pastoralists living in the agricultural *douars*.

Recently, the sedentarization of nomadic pastoralists has taken a new twist. Formerly they abandoned their pastoral space just to settle in 'tent towns' at the gates of small urban centers (Tag 2003), as it was the case during the droughts in the late 1970s and early 1980s. The contemporary type of sedentarization, on the contrary, occurs on parts of the steppe that were made arable. Generally this land is tilled by tractors and sown with wheat and barley. Then the pastoralist builds a house in vicinity to the tent. In the course of time, the tent, in a process of decay, is more and more reduced to a simple *oucha*, a hut made of braided mats of *alfa*, sheltering some sheep, goats and poultry. The new permanent mansion is now the family's safehouse. This new kind of hybrid habitat essentially mushrooms alongside the main herding tracks that have multiplied in the recent past in eastern Morocco. Such a type of sedentarization represents the informal appropriation of a portion of the collective lands. It may be the result of an agreement between household heads belonging to the same tribal fraction or tribe, having an ancestral right of usage recognized by the *jemaa* of the tribe. The size of these farms, which result from the transformation of rangeland into cropland, differ from one person to the other, depending on his social and economic rank within the tribe. According to oral reports, the most influential persons have acquired large farms exceeding 20 ha.

Additionally these matter-of-fact appropriations have generally taken place where once good pastoral meadows of *artemisia* existed, such as around Dayet Tinzi in the north, and on the occidental Dahra of the Oulad el Haj. Hence, in resorting to these informal appropriations, the pastoral communities of the high plateaus strive to secure the permanence of their existence in that area, which is supported by the dynamics of the shadow economy in the vicinity of the transfrontier spaces with Melilla, occupied by Spain, and with Algeria. Simultaneously, these communities aim at conserving their rights to the collective pastures. Moreover, since the early 1990s a new phenomenon has arisen on the northern boundaries of the high plateaus, particularly concerning the zone of Tafrata, where agro-pastoral farms in pastoral communities are now being purchased by the urban bourgeoisie or the newly rich who launder their money in the real estate business.

The pastoral landscape has thus been profoundly transformed. On the eastern, northern and western boundaries the rangelands are making room for agricultural farms equipped with houses and wells. On the high plateaus, three decades ago only herds of sheep and goat were encountered, while the tents were scattered in the *alfa* or in the depressions protected from the winds. Nowadays 'hybrid habitats' are emerging amidst the *alfa*, as well as in the middle of cereal fields and olive-plantations, irrigated by water from artificial wells, as the nomadic pastoralist of the past progressively turns agro-pastoralist.

For a long time the steppes of the eastern Moroccan plateaus were lacking an urban center with the exception of the city of Oujda, the regional capital of the entire eastern part of Morocco. Since the late 1970s, there has been a blossom and growth of a certain number of urban centers, which form an urban corridor (Map 5-1). These urban centers have polarized the respective spaces by reinforcing their function of local or regional *souks*, and in diversifying their tertiary activities serving the pastoral and agro-pastoral communities of the region (cf. Tag 1984).

Before the 1970s, access to the high plateaus was problematic. The means of transport were pack animals and dromedaries. Water holes were scarce and little known. Insecurity was widespread, as was the menacing hyena. But since the inception of the territorial political administration and the communal partition, considerable measures have been implemented, such as tracks, hill barrages and individual or collective motorized transportation. Henceforth the high plateaus are opened for circulation and exchange and even international tourism has begun to touch the plateaus. Spanish tourists are prominent.

Social Transformations

Already in the 1980s the dissolution of nomadism on the steppes of the eastern Moroccan plateaus was observed (cf. Tag 1984). This dissolution has affected the very foundations of nomadic social structures as much as it has their way of life. But the transformations of the pastoral society have continued during the 1990s.[2] Three aspects are crucial:

1. School enrollment: A person told me: "Our parents and our ancestors have squandered us in making us run after sheep and other animals." In effect, the nomadic pastoralists of this region have stayed illiterate for a long time. The isolation in the steppe protected them from the administration's demands and left them at distance from any education. During the last quarter of the past century, things have changed. The heads of wealthy families have become conscious of the need for school and education. Hence, in purchasing houses in the region's different urban centers they have been able to enroll some of their sons and daughters in school. These same families, thanks to their connections to individuals well-placed in the administration, have been able to place their children with university degrees in different positions within the public sector, like local administration, military, and other sectors. Others have supported their children in their quest for employment in the private sector, helping them to become advocates, doctors or employees in private enterprises.

2. Migration: The intra-territorial movements motivated by the search for pastures are disappearing. This can be explained by endogenous factors tied to the changes of the bygone pastoral way of life. A fair amount of the pastoral population is part of the rural exodus and others contribute to the emigration towards foreign countries. Those that take part in the rural-urban migration are among the most needy: They have not resisted the natural catastrophes of droughts. They exert themselves as laborers in the construction sector or as porters with pushed or pulled wagons. International emigration, on the contrary, has attracted sons of wealthy families for which emigration – for example to Spain, France, Belgium and Italy – is a social sign of success. Some of these emigrants actively participate in the spatial transformations of the region by purchasing collective pastoral terrains, investing in agricultural mechanization, and by the amelioration of the livestock. The migratory phenomenon inscribes itself in the frame of the population's general mobility, enhanced by the effects of globalization.

3. Social differentiations: The pastoralists of the region have been subject to multifaceted processes of social differentiation. Today, there are large differences among them, especially in social status, vulnerability, and economic strategies. I propose the subsequent classification: The very poor and vulnerable pastoralists on the one hand and the rich stock producers on the other hand. These latter ones can be sub-

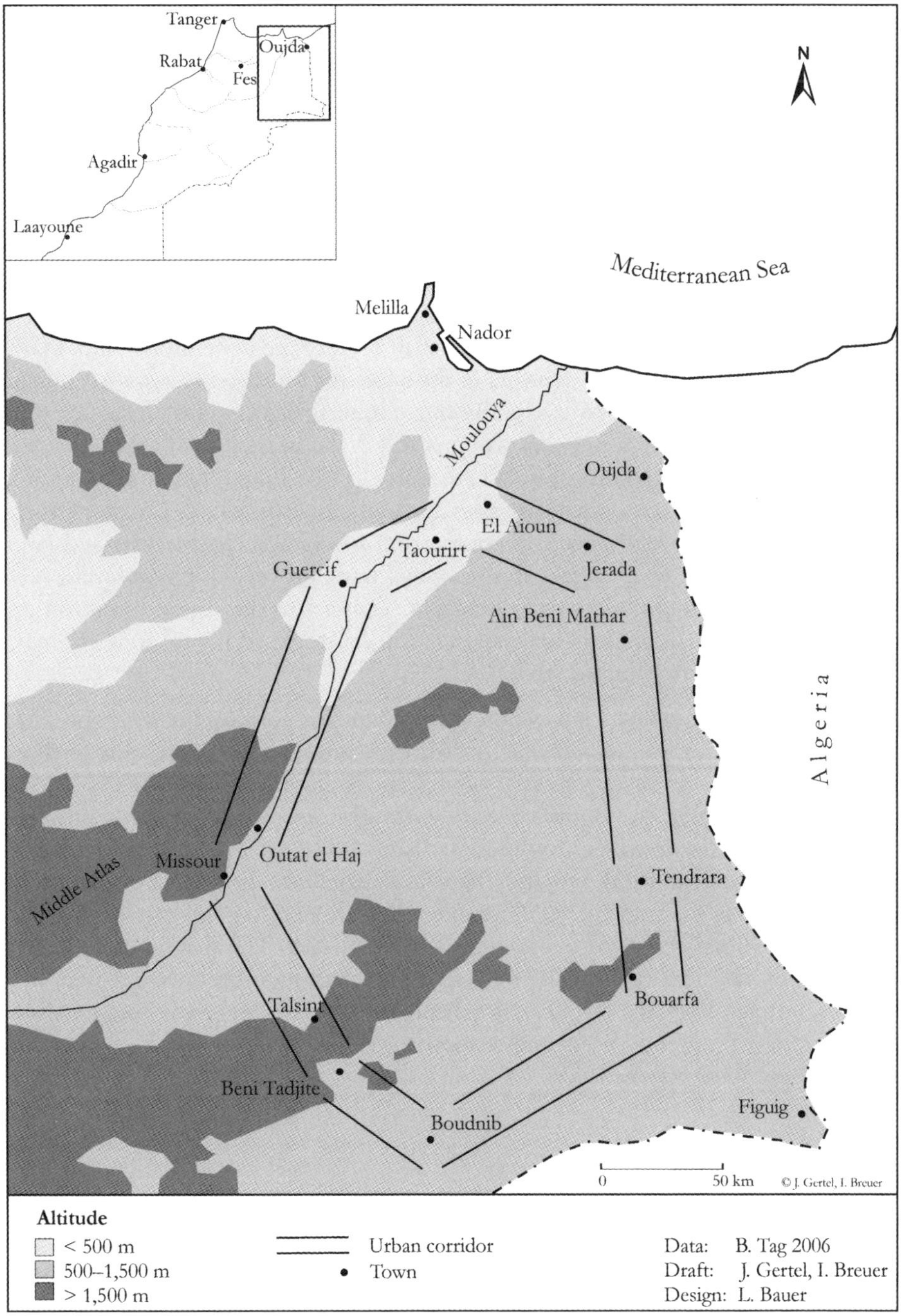

Map 5-1: The Urban Corridor of the High Plateaus

divided in three groups, according to their kind of economic activity: (a) Large livestock producers. They possess herds of great size nearing up to 500 sheep or goats, and employ more than one shepherd. Livestock production is normally their only activity. They are strongly market-oriented, which is also reflected in their breeding techniques: the animals which are destined to be commercialized are separated from those reserved for reproduction. The latter, generally ewes and female goats, as well as the best progenitors, constitute the basic capital. These livestock producers are well-equipped with 'modern' stock-breeding equipment, such as metallic cribs and drinking troughs, trucks furnished with motor pumps and cisterns for water supply and even motorcycles, and reserves of supplementary fodder. They often own real estate in the urban centers and agricultural land. Many of them also possess financial deposits in the bank. (b) The agro-pastoralists pursue strategies of economic diversification by integrating agriculture into livestock production. On the high plateaus, this group is still at the beginning of its appearance, due to the lack of irrigation water. Where water can be found, like in the corridor Guercif-Oujda, agro-pastoralists are already firmly established. (c) Livestock speculators form a very small group in comparison to the agro-pastoralists and large livestock producers. They engage in the animal trade on very different levels, such as wholesale and retail traders or as butchers. They roam the region's *souks* in order to engage in small or large-scale speculation on stock, trying to draw the maximum profit from price fluctuations.

Some wealthy stock owners have entered the country's political life recently. They join political parties, sometimes due to their own interest, sometimes due to those of a party that sees in them potential candidates during the communal and legislative elections. Hence, pastoralists are nowadays present in the region's different communal councils, or even move up to parliament. Here it would be interesting to ask whether their political activities benefit pastoralists. In addition it must be noted that the system of animal husbandry has also undergone certain changes. Finding a qualified shepherd has become a difficult task. The shepherds, formerly paid in kind, are now reimbursed by a monetary and monthly salary, based on a written contract. Even the term *berger* (shepherd) bearing a pejorative connotation, is less and less frequently used, and is substituted by the term *gérant de troupeau*, which means 'livestock manager'.

Market Dynamics

The role of pastoral livestock production for the regional economy cannot be overestimated: it is its very basis. While the lower Moulouya region, directly north of the steppes, is known to be a modern agricultural center important on the national agenda, the steppes of the eastern Moroccan plateaus almost entirely depend on pastoral production. One of the most important transformations concerns the meat and fodder markets. Fodder, especially barley, has recently been adopted by the pastoralists to ameliorate the quality of the stock destined to be sold on *souks*, while the herd mobility is being reduced. The majority of pastoralists have also integrated cattle. The calves are destined for sale on the markets and the cows for the family's milk supply. Cow milk increasingly replaces sheep or goat milk, which in turn will be provided to the young animals in order to obtain a good stock quality. The introduction of cattle – sometimes of excellent breed, like the *Pie Noir* – has been facilitated by their introduction to Algeria during the early 1990s at the time when the mad cow disease let prices fall in Europe. This is an effect of globalization that reaches up to the steppes of the eastern Moroccan plateaus. In the same transfrontier context, the Algerian *Oulad Jellal* sheep breed has rapidly expanded across eastern Morocco. Among some pastoralists of the Ain Beni Mathar region, 80% of the herds are already *Oulad Jellal*, and this breed does not cease to gain terrain. Economically, due to its weight it is more profitable for the butcher than the local *Beni Guil* breed, and it is consequently more and more appreciated by livestock producers.

Further, camel keeping has experienced a revitalization. Since the integration of the Western Sahara to the national territory, this has become a promising market gap, since this region's population strongly demands camel meat. Thus, camel raising will experience a renaissance in all the semi-arid spaces of Morocco. In the eastern Moroccan steppes, this activity reappears with certain wealthy pastoralist families of the Ahl Teggour tribe south of Outat el Haj. Furthermore, goat and camel meat are increasingly demanded in the cities, as is camel milk, renowned for its excellent quality. Hence, the rearing of dromedaries constitutes a promising and profitable niche in the market (cf. Tag 2003).

Conclusion

Recent processes induced by endogenous and exogenous factors have fundamentally transformed the traditional nomadic pastoralism in the steppes of the eastern Moroccan high plateaus. The locational significance of the region still promises more important changes, namely in the context of further regional integration. Here, three potential dimensions can be cited: A Saharan dimension of integration towards Saharan and Sub-Saharan countries; a Maghreb dimension if the contentious relationship between Algeria and Morocco can be appeased; a Mediterranean dimension given that the region of the Oriental is open to that space by means of two airports, namely Oujda Angad and Farkhana, one maritime port (Nador), and the small European port of Melilla. These dimensions will certainly reinforce the impact of globalization on pastoralism within the region.

Notes

(1) Between 1995 and 2000 I have participated as consultant in the realization of the SDAR (Schéma Directeur d'Aménagement Régional) of the Oriental's economic region.

–6–

Pastoralism and Institutional Change in the Oriental

Mohamed Mahdi

Pastoral communities of the Oriental[1] are known for their characteristic nomadic life. They are socially structured in nested ethnic groups – tribes, fractions, lineages and *douars*, 'the nomadic encampments'. The *douar* is a basic unit of transhumance and of social organization in nomadic life. Every nomadic camp is composed of a central nucleus, represented by a large and strong tent (khaima kbira), and satellite tents. The latter might be inhabited by members of the grand tent's lineage, their clients and servants.[2] The pastoral territory is subdivided in *woulfs*, literally meaning "habitual place". It represents an ethnic group's area of transhumance, including its summer and winter pasturelands, its farmlands and its water resources. The customary law, *ᶜurf*, regulates the usage of the *woulfs*. Finally, raising livestock depends on mobility, in other words nomadic horizontal transhumance between summer and winter pastures.

This 'reference model' of nomadic society, described by the three pillars society, territory and technology can in our days no longer be found in such a pristine state, since numerous factors have intervened to transform it. Moreover, while the literature on nomadism evinces that the mode of nomadic life has never been static, but rather ever evolving and mutative (Mahdi 2002a), today nomadism in its generic form no longer seems viable. The pastoralists themselves search for alternatives that are better adapted to new socio-economic, environmental and other conditions. Hence, in the eastern Moroccan high plateaus forms of nomadism coexist or mix with new, more intensive systems of livestock management. Subsistence animal husbandry based on an economy for survival and conducted by poor farmers persists next to an outsourced livestock management performed on behalf of some rich entrepreneurs. Pastoralism in the Oriental is thus increasingly complex. Consequently the classification of livestock related enterprises and of pastoralists is an extremely arduous task.[3]

In the following, three dimensions of transformations are investigated:[4] social changes, such as the processes of sedentarization and migration; ethnic and institutional transformations linked to the introduction of cooperatives; and finally technological transformations, such as the adoption of feedlots, trucks and supplementary fodder. In conclusion this paper argues that the subsequent reorganization of pastoral institutions fostered new inequalities, where the poorest are increasingly marginalized.

Social and Demographic Transformations

Mobility and nomadism have their antipodes, namely immobility and sedentarization. In the Oriental, sedentarization and temporary sedentariness of nomads are the result of an ancient process, one that has increased its scope in the last three decades. This is due to recurrent droughts and the consecutive reduction of herd sizes. The literature on the Oriental abounds with indications of this phenomenon. Due to accommodativeness or compulsion, sedentarization has taken different forms: the residency in a tent or in a hut on the pasture, the habitation in a house in an urban center or in a tent or hut at its periphery.

The extent of sedentarization becomes more evident by analysis of the statistical data available regarding the changes that have become apparent in the type of habitat and in the attitude of livestock owners towards herd mobility and nomadism. According to a survey[5], started in 1989 and completed by another in 1991, 42% of the questioned household heads possessed a house in the city, 40% lived in permanent houses, whereas 60% dwelt in tents or *oucha*, huts made of braided mats of *alfa*. But one must not be deceived, merely possessing a house in town and even settling or attaching oneself to the rangelands, does not exclude the herd's mobility. Rather the propensity towards settling can frequently be explained by the desire to improve living conditions on the rangelands, the wish to dwell in town or to open up new livelihood options.[6] The same survey revealed that 56% of the pastoralists had abandoned transhumance, while 44% continued to transhume at the time of enquiry. The abandonment of transhumance has been progressive and constantly rising. The livestock owners who had forsaken transhumance did not represent more than seven percent before the 1960s. The abandoning of spatial mobility involved 22% of them between 1960 and 1970 and 28% between 1970 and 1980. It comprised 35% between 1980 and 1991 and seems to further increase.

Nomadism is increasingly motorized, the truck is replacing the camel as means of transport. This motorization evolved in recent years and is a phenomenon that overthrew the guiding principles of traditional social relationships. It also altered the norms which structured nomadic encampments: progressively, trade relations substitute traditional forms of exchange and solidarity.[7] In effect this means that nomadism has become the privilege of those who have the means.

Changes in lifestyle, in housing conditions and in herd mobility favor quite remarkable demographic dynamics, as well as urbanization in the Oriental. New territorial and administrative units appear in the landscape. Small urban centers, like Ain Beni Mathar and Bouarfa, attain the status of municipality. Other urban centers, like Tendrara, developed, while some new rural communes emerged after the administrative reorganization of 1992 (cf. Map 6-1). These new urban agglomerations

Table 6-1: Eastern Moroccan Highlands – Population

Municipalities (m), urban centers (c) and rural communes	1971	1982	1994	2004
Bouarfa (m)	6,181	13,471	19,631	25,947
Ain Beni Mathar (m)	3,356	5,176	10,532	13,526
Tendrara (c)	16,232	14,186	11,328	12,057
Abou Lakhal	–	–	1,862	1,497
Beni Guil	–	–	9,014	9,059
Maatarka	–	–	8,623	8,030
Tendrara	–	–	–	12,057
El Atef	6,124	6,665	2,186	2,471
Oulad Mhammed	--	--	3,487	2,174
Beni Mathar	5,673	6,886	5,207	7,089
Merija	6,869	7,015	2,198	2,841
Oulad Ghziyel	–	–	5,596	6,488
Oulad Sidi Abdelhakem	2,905	4,855	3,001	2,995

Source: RGPH official census data.
Remarks: Administrative borders have been subject to changes between different census years. Thus, comparison between population figures between years and between administrative units is sometimes not possible.

Table 6-2: Eastern Moroccan Highlands – Annual Population Growth

Municipalities (m), urban centers (c) and rural communes	1971 to 82 (%)	1982 to 94 (%)	1994 to 2004 (%)
Bouarfa (m)	4.5	2.4	2.8
Ain Beni Mathar (m)	2.9	3.9	2.5
Tendrara (c)	- 1.2	4.6	0.6
Abou Lakhal	–	–	- 2.2
Beni Guil	–	–	0
Maatarka	–	–	- 0.7
El Atef	0.7	–	1.2
Oulad Mhammed	–	–	- 4.6
Beni Mathar	1.5	- 2.4	3.1
Merija	0.2	- 2.6	2.6
Oulad Ghziyel	–	–	1.5
Oulad Sidi Abdelhakem	3.3	- 4.7	0

Source: RGPH official census data.
Remarks: Administrative borders have been subject to changes between different census years. Thus, comparison between population figures between years and between administrative units is sometimes not possible.

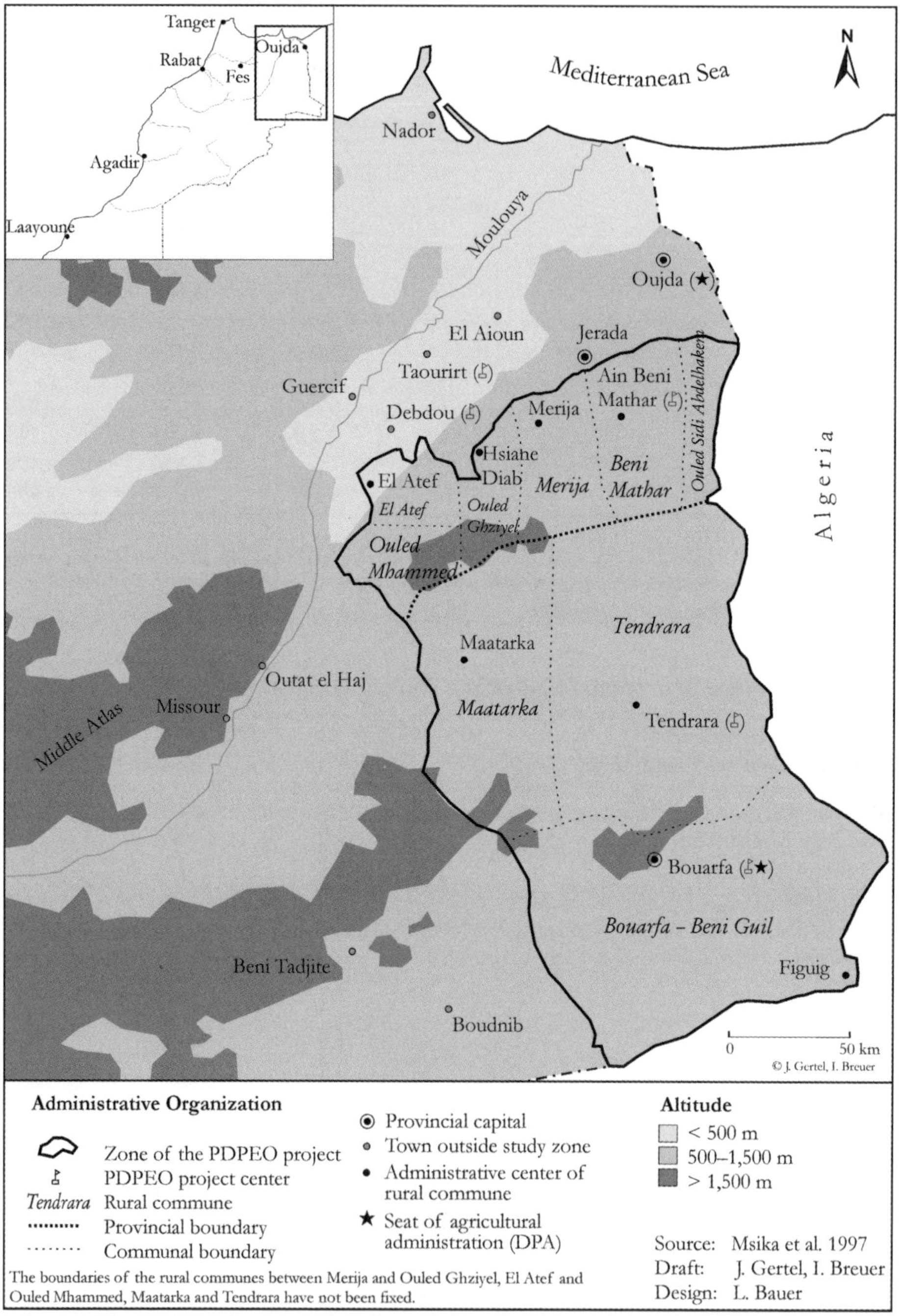

Map 6-1: The PDPEO Zone – Administrative Setting

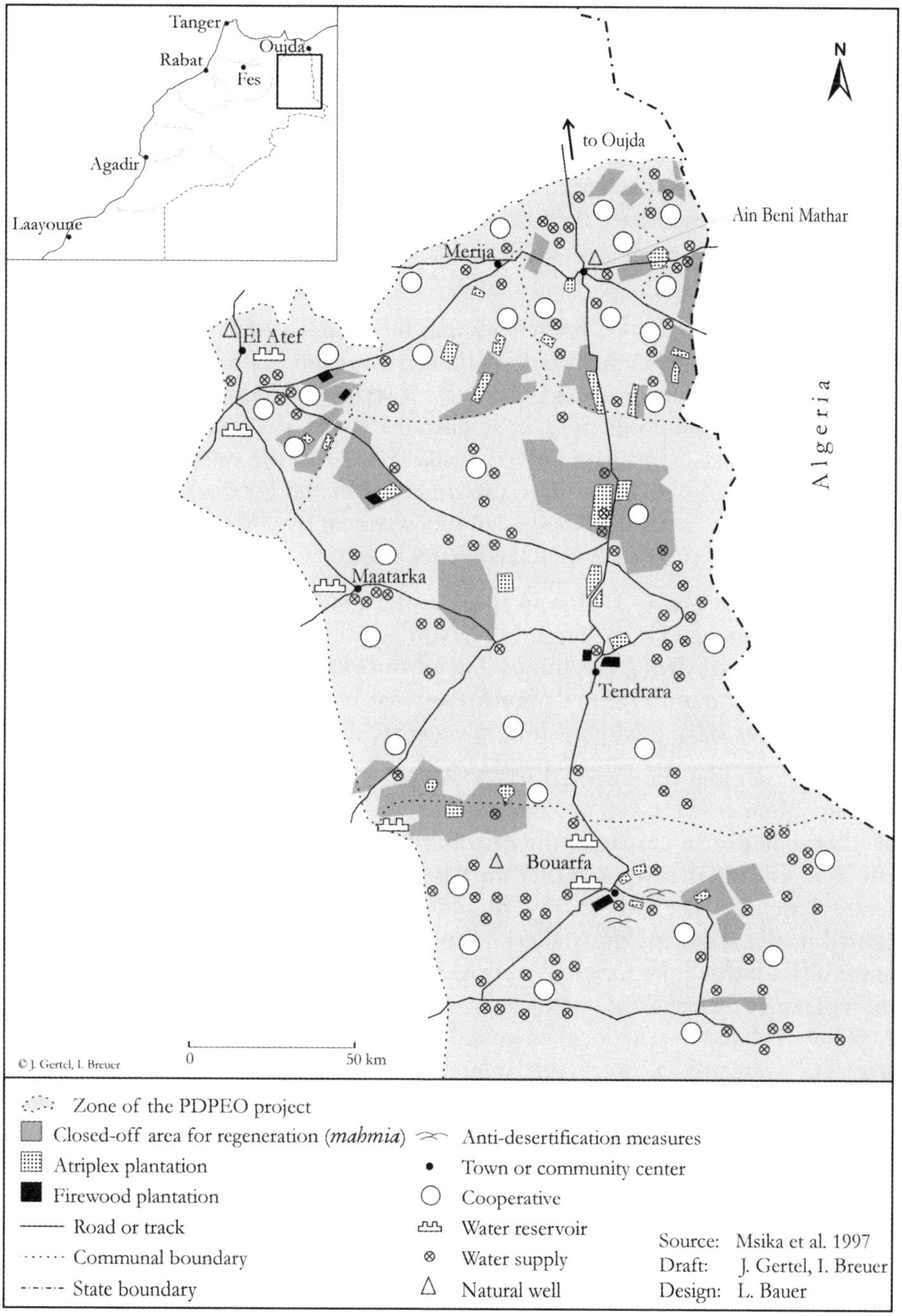

Map 6-2: Geography of Interventions in the PDPEO Zone

attract livestock holders and accelerate the process of rural-urban migration and sedentarization in their vicinity. A provisional analysis of official census data[8] supports the hypothesis of sedentarization and rural-urban migration as principal explanatory factors for the demographic evolution of the Oriental. The annual average population growth rate in the municipalities and urban centers significantly exceeds the national average, while the majority of rural communes reflect an annual average growth rate well below this.[9] Generally, the rural communes are being depopulated. In contrast, municipalities and urban centers grow at rapid rates (Tables 6-1 and 6-2).

The movement of livestock holders leaving their rural communes to settle in the emerging urban centers – such as Bouarfa and Ain Beni Mathar and to a lesser degree Tendrara and Merija – might be either a strategy to adopt better performing systems for livestock production, or a consecutive step of retreating to a city, or its periphery, to engage in urban animal husbandry, or in the extreme case to adopt some other activity after the utter abandonment of the livestock business. These mobilities are also marked by a population movement from south to north, notably in consequence of bankruptcies in the livestock sector.

The population's social decline in these centers, stemming from sedentarization and the rural exodus, is the origin of certain social upheavals of the past months. Frustrated by their living conditions, the inhabitants of some agglomerations have manifested their discontent in demonstrations, at Bouarfa in September and at Talsint in October 2005, predominantly to vindicate their right to employment.

In the past decades the Oriental has witnessed parallel to sedentarization the unfolding of immense internal and international migration. This migration constitutes another indicator of change in the pastoralists' lives. It also indicates the extent of efforts to find alternatives to buffer the crisis in which the pastoral economy is currently trapped. The ERP survey (Enquête Rurale Participative – PDPEO II), carried out in 2002, in which 4,064 adherents to ten pastoral cooperatives in the north and south of the Oriental were surveyed, has permitted data collection on the migratory behavior of families. It was ascertained that these families count 2,088 internal and international emigrants. Hence in 52% of the families there is at least one emigrant. 57% of this migration is international, internal migration constitutes the rest (cf. Table 6-3). Whereas the internal migration is in most cases undertaken by livestock owners or their family members originating from rural communes in the south, the international migration is far more frequent among people from the north. Emigration towards foreign countries depends on the relative affluence of each zone. Its weakness in the south is an indicator of more accentuated poverty and of the unavailability of emigration networks to foreign countries.

Table 6-3: Eastern Moroccan Highlands – Internal and International Emigration

Community	Pastoralists questioned	Emigrants	Ratio	Distribution of Emigrants (%)	
	(1)	(2)	(3)	Morocco	Abroad
Beni Guil / Abou Lakhal	1,574	423	27	81.8	18.2
Tendrara / Maatarka	1,089	496	46	68.3	31.7
Merija / Oulad Ghziyel	286	275	96	7.6	92.4
Ain Beni Mathar	211	179	85	17.8	83.2
Oulad Sidi Abdelhakem	102	124	122	25.0	75.0
El Atef / Oulad Mhammed	802	591	73	20.6	79.4
Total	4,064	2,088	–	42.6	57.4

Source: Enquête Rurale Participative – PDPEO II
Remarks: (1) The 4,064 pastoralists questioned include adherents to ten pastoral cooperatives.
(2) Total number of emigrants in the households of surveyed pastoralists.
(3) Ratio = Number of emigrants in the households of pastoralists per 100 pastoralists interrogated.

The destination of internal migration can be either located in the research area itself, or in the northern part of the Oriental, from Jerada to Berkane and from Oujda to Guercif, and finally in the cities outside of the region. It is noteworthy that 32% of the internal emigrants settle in the center of their communes and 35.4% in the center of neighboring communes. Nearly 31% head away from the survey area without abandoning however the region of the Oriental. The population thus leaves the countryside to settle in the urban centers. In effect, a fair percentage of these emigrants settles in the surrounding and in the immediate periphery of urban centers at Oujda, Bouarfa, Jerada, Figuig, Tendrara and Ain Beni Mathar.

The analysis of migration patterns by ERP reveals that the use of a tent close to an urban center is often connected to the modesty of herd size: it concerns predominantly stock owners who possess less than 60 sheep. This form of migration may simultaneously be motivated by the quest for employment. The majority of the internal migrants practice professions such as day laborer (48%), shepherd or stableman (24%) and agricultural laborer (6%).

93% of international migration is oriented towards southern Spain, where the emigrants seek employment as agricultural laborers. This emigration existed in the Oriental before 1990, but it did not involve more than 5% of the emigrants registered in the census. It was only after that date, that international emigration increased significantly to 23% of emigrants between 1990 and 1995 and 68% between 1996 and 2000. Emigrants do not cut their ties with their home community. Rather many of them regularly move between Spain and their place of origin. They virtually are 'both in Spain and on the pasturelands'. The ideal expressed by some of them consists of sustaining this coming-and-going in such a manner that they can take care of their herds and farms. Emigration thus allows them to retain the herds and the tent as integral parts in their lives, even if they spend less time with them. Of course, some of these emigrants also lease out their herds or they entrust them to a family member.

Ethnic and Cooperative Administration of Pastures

The administration of pastures is traditionally founded on an ancestral ethno-territorial organization. It is also based on a tribal consensus regarding the mutual recognition of groups on the given rangelands, called *woulfs*. The pastoral cooperatives, either based on ethnic lineages or ethnic territories, like the new framework of pastoral organization and administration of the rangelands initiated by the Project of Pastoral Development and Livestock in the Oriental (PDPEO), have in their conception taken into account this traditional organization.[10] The cooperative based on ethnic lineages (coopérative ethno-lignagère) appears to be an upgrade of a modern institution, the cooperative, combined with traditional institutions, the 'socio-spatial-ethnic units'.

The designers of the *coopératives ethno-lignagères* in collaboration with the pastoralists, have based admission to the cooperative on the following conditions: adherence to the ethnic group; holding the right to use collective pastures; practicing pastoralism as principal means of livelihood; and residing in the rural community where the cooperative is located. The cooperative's role is – among others – to serve as the administration's negotiator regarding the range management. It also has to guarantee the amelioration of pastoral land, as well as the provisioning and distribution of subsidized and commercial fodder to its members.

In this manner the pastoral cooperatives of the Oriental, in agreement with the project's administration, have closed off certain pastures for regeneration (mahmia) (cf. Map 6-2). These pastures are generally reopened after they regenerated.[11] In return the cooperatives receive fodder for the stock. The use of pasturelands placed under *mahmia* and the distribution of subsidized fodder illustrate the change in the nature of the groups' relationship to their pastures. For, after the period of regeneration, these rangelands are in principle only open to members of the cooperative. Moreover, only members receive the compensatory food items. Those, who are not registered in the cooperative receive only a part of the distributed fodder. Each livestock producer – even a non-member of the cooperative – can use the pastures, but he has to pay the double price for the use right, compared to the cooperative member.

The creation of the cooperative thus substitutes the traditional rights of access with a membership status. At the same time, the practice of closing off areas for regeneration marks the passage from the understanding of pastoral territory in terms of *woulfs* to its new perception as *mahmia*. These two notions are ruled by different logics. The *mahmia* thus has become exclusive, since access is now in principle bound to formal membership. Legitimate claims are thus not only based on ethnic affiliations, but also depend on the shares a member holds of the cooperative's capital. The quantities of subsidized fodder allocated to each member will be calculated on the respective basis of these shares.[12]

It is not surprising to detect numerous violations of closed-off *mahmia*-land. Often the violators are from the same ethnic collective or from bordering ethnic units. These latter ones, in particular, reclaim ancestral access rights to the rangelands just after its closure. Appeals for a pastoral police and the creation of a continuous surveillance system become insistent in order to protect the *mahmias* and the privileges with which the cooperatives endow their members.

This membership status lends privileges which some have not hesitated to transform into a system of land rent. The adherent will use his access right to the *mahmias* either directly by letting his herd graze there, or indirectly by entrusting his herd to a shepherd or an associate. Or finally he may make use of his right to access regenerated sites by renting it out to a non-member, who pays a sum for each sheep in advance for one month of pasture.[13]

The cooperative has thus become central for the administration of rangelands and progressively substitutes traditional customary institutions. The final report of the PDPEO project (DPA 2002) recognizes these cooperatives for their indispensable partnership in the administration and development of the pastoral space and for the distinct roles they play, for instance, in the delineation of regeneration sites, the nomination and remuneration of guards and the punishment of violators. These factors constitute a new relationship between pastoral communities and their territories.

The new strategy of cooperative development does not represent a complete and effective answer to 'the tragedy of the commons', because it does not exclude individual and individualistic strategies. On the contrary, the process of individual appropriation of pastures through creation of plots with rain-fed cultivation has increased since 1990. The surface of these areas is estimated to be 150,000 ha of which 147,000 can be classified as former collective pasture. These lands were individually appropriated with or without consent of the local authorities.

Pastoralists and Livestock of a New Type

The changes that have affected the socio-spatial components of nomadism, have also modified its technical dimension. In this zone of ancient pastoral tradition, livestock raising, which remains the foundation of the economy, is presently conducted in 'systems' which break with nomadism in its 'original' or old-fashioned form. More intensive and strongly market-integrated production systems developed and offer the possibility to safeguard pastoral livelihoods, even if only provisionally. It has to be added that government assistance, provided in the framework of the PDPEO project, has boosted these unfolding systems of livestock production and also accentuated their intensification. The pastoral cooperatives have taken the lead in supplying the pastoralists with fodder at competitive prices.

This development has certainly enabled some nomads to break the vicious circle of poverty, where each drought meant a massive reduction of herd size. Nonetheless it has subjugated livestock production to the yoke of financial capital while the pastoralists' livelihood depends on his capacity to invest in the purchase of food, stall construction and the acquisition of trucks to transport the animals, water, etc. Those endowed with the necessary financial means succeed, the poor succumb. The transformations in the pastoral livestock production system entail progressively widening socio-economic disparities.

The studies devoted to the classification of livestock and pastoralists are full of indications about the polarization of the animal production system and the concomitant hierarchies. A study realized by the INRA (Institut National de la Recherche Agronomique) has identified three types of livestock production in the Oriental.[14] First, the 'subsistence producers' represent 67% of the pastoralists queried, but only hold 19% of animals. They raise small herds, are often in pecuniary difficulties and only survive by means of the pluri-activity of family members. Second, the pastoralists who live in 'agricultural units', with 19% of livestock producers and 16% of the animals. Here livestock raising is more a way of life and a means of living than just a profitable capital for monetarization. Finally, some pastoralists raise their livestock in 'units of speculation', directed by a rationality of enterprise. They represent 14% of all stockbreeders and own 65% of animals. Here the production's objectives are the maximization of profits and the quest for greatest profitability of capital investments.

Another classification proposed in the framework of the ERP differentiates the pastoralists according to their relationship to the pastures and offers the possibility to include the evaluation of migration. A first type of pastoralists is constituted by those who still work 'on the pasture grounds'. These can be subdivided in three groups: nomadic people, about to disappear, pastoralists, settled in town or on the rangeland, but carrying out mobile livestock production on the region's pastures and, finally, sedentarized livestock producers using pastures and also stalls. The second type is made up of 'peri-urban livestock producers', who left the pasture and settled in a city's periphery. Among these it is possible to distinguish between those owning herds of small size, settled close to the center of their community, and those in bankruptcy, who only hold less than five animals and work as day laborers. Third, there are the 'urban livestock holders', who employ feedlots. Generally they are affluent, have a permanent job and have owned houses for years. Fourth, the stockbreeders that are abroad, especially in Spain, and who migrate to and from Morocco. Finally those, who abandoned livestock production and try to obtain employment as shepherds or stablemen. They await an opportunity to build up a herd and to return to the business.

From the point of view of their relative importance, the first group, living on the pasture, represent 48.7%. Those of the city – that is the urban and peri-urban livestock producers and those who have quit the activity – constitute 21.8%. Those abroad make up 29.5%. Thus the pastoralists of the Oriental do not compose a monolithic and homogeneous group. They rather differ in various regards, concerning their livestock capital, their relationship to pastoral mobility, their attachment to urban centers and their strategies in the livestock production sector.

Conclusion

Numerous transformations have been affecting pastoralism in the Oriental for several decades. This paper focuses on those institutional changes that touch the social, spatial and technical dimensions of this traditional way of life. One objective is to demonstrate that these transformations have a historical dimension. At times the transformations are spontaneous, at times provoked. They parallel forms of change affecting pastoralism in other Moroccan localities and even in other regions, at least in the Maghreb countries. Pastoralism as an activity of livestock production, bound to the pasture grounds and based on the mobility of herds, has increasingly been abandoned in favor of fixed, intensive and more profitable forms of animal husbandry.

These transformations collaterally effect changes in the pastoral communities, their institutions, and their territories. More than ever, this inseparably links the question of pastoralism's future to that of the development of the Oriental as a whole. The second phase of PDPEO declared the reduction of poverty as its primary objective, but the transformations of pastoralism entail new social hierarchies, new forms of inequalities. They have led to the birth of new social groups, living more and more precariously, especially the women, the young with or without education, and those excluded from livestock keeping. All of them inflate the army of the unemployed that vegetate in the new urban centers and at their peripheries.
Regarding the development of livestock and pasture, the massive import of animal fodder has fostered combinations of livestock production on rangelands and feedlot employment. The first gives room for the niche of nomadism, for those who have the means. The fodder import has also helped to keep this economic activity alive. The development of the Oriental will call for numerous arbitrages between the development of livestock production and pastures on the one hand and the promotion of new economic and social activities on the other, to directly provide solutions for the population's many expectations and to ease the latent tensions in the region.

Notes

(1) The Oriental is here defined as the zone limited in the north by the Oujda and Debdou mountains, in the east by the Moroccan-Algerian border, in the west by the Rekkam plateau and the mountain chain of the High Atlas, and in the south by the pre-Saharan zone. It comprises the communes of Ain Beni Mathar, Merija, Oulad Ghziyel, and Oulad Sidi Abdelhakem (province of Jerada), El Atef, Oulad Mhammed (province of Taourirt) and Tendrara, Maatarka and Beni Guil (province of Figuig). This zone equals the area of the PDPEO-project (cf. Map 6-1) with a surface of about 32,000 km².

(2) Rachik (2002).

(3) The end of nomadism was declared on many occasions, but each time the account was rendered that nomadism continues to exist, only in other forms. Cf. Rachik (2000), and this vol.

(4) "The change of the production system cannot be considered in isolation. Rather it has to integrate social, institutional and political factors that could be regarded as mediators of the relationship between the population and environment." (translated from French), in M. Picquet et. al. (2004), 28.

(5) The provincial directive Board of Agriculture conducted a 'livestock survey' in 1989, which questioned 8,938 household heads, fanned out in six rural communes: El Atef, Merija, A.B. Mathar, Oulad Sidi Abdelhakem, Tendrara, Bouarfa. This survey was completed by the IAV *(Institut Agronomique et Vétérinaire Hassan II, Rabat)* in 1991 focusing on 1,500 household heads in the same rural communes, see Hammoudi et al. (1992).

(6) The best management of livestock raising opportunities consists of arranging a double habitat, one on the rangelands and the other in town, where two households are installed. One takes care of the flocks, the other accommodates itself in town and serves as a reception centre for family members, who enroll in school, work in town, or come there because of the market, the hospital and for the other services lacking on the rangelands.

(7) See Rachik (2000).

(8) For instance the RGPH (Recensement Général de la Population et de l'Habitat) of 1971, 1982, 1994 and 2004 on municipal, urban and rural-communal level.

(9) Between the two RGPH 1994 and 2004, the annual average growth rates are 0.6% for the rural population, 2.1% for the urban population and 1.4% for the entire Moroccan population. The synchronization of data from different surveys demonstrates that in addition to natural population growth, there is a simultaneous effect of sedentarization and of a rural exodus.

(10) The originality of the PDPEO was the initiation of pastoral cooperatives, which furnish the organizational frame for livestock breeders in the Oriental. So far 44 co-operatives have been created.

(11) The project initially envisioned the regeneration of 750,000 ha, 61% (470,000 ha) of which have been realised (cf. DPA-Bouarfa 2002).

(12) Certain cooperatives have tried to re-establish some equality in prescribing 10 to 15% of the fodder allocations to the poor and to non-livestock-producing members of the ethnic group (1 to 2 sacs of 80 kg). Cf. Harrach (1997).

(13) Harrach (1997).

(14) Differentiated by beginning with functional variables (nutritional status, the main source of income) and structural variables, notably the herd size and the means of transport.

–7–

Trading Livestock: Eastern Moroccan Sheep Meat Commodity Chains

Mohamed Khalil[1]

This paper analyzes sheep and sheep meat commodity chains linking the production sites on the eastern Moroccan high plateaus with the consumption centers, mainly the major cities of the country. Of specific interest are the strategies of the actors at different levels of the sheep commodity chain – like pastoralists, and different types of livestock traders. It will be shown that the Moroccan sheep commodity chain is far from homogeneous, comprising different interrelated economic circuits, in many of which certain groups of pastoralists take actively part. The study is based on a survey conducted on the eastern Moroccan high plateaus in 1996/97. Fieldwork included interviews with several hundreds of livestock traders (maquignons) of whom 165 come from outside the study area to buy livestock. Moreover, the survey included 49 livestock producers with different herd sizes (large, middle, small) and two important local markets (souks) of the region, where data on transactions were regularly collected.

The Eastern Moroccan High Plateaus

The study area covers the center and the south of the eastern Moroccan high plateaus. It comprises the nine rural communes (communes rurales) affected by the PDPEO[2] development project (cf. Mahdi, this vol., Map 6-1). Its total surface area is 32,000 km², with a population of 86,993 (1994), 53.8% of whom are considered as rural. The study area is home of about one million sheep. Until today, the region has been among the most important sheep meat producing areas in Morocco, total sheep production accounting for 400,000 head per year. Simultaneously the study area is characterized by large exports to other Moroccan regions (900,000 head per annum) and weak local consumption (32,000 head per annum). The substantial difference between production and exports is only to be explained by movement of animals out of Algeria into the region. The study area thus plays an important role as a initial reception and trading point for Algerian animals.

The two main *souks* of the zone are Tendrara and Ain Beni Mathar. These are very rudimentarily organized and weakly equipped markets, where data on trade are sparse. The *souk* of Taourirt assures an important link in the network between the region and the rest of the country.

Within the study area, sheep production is very variable and depends on the climatic conditions which display substantial inter-annual differences in precipitation. The production system in this environment of strong climatic fluctuations is extensive, marked by seasonal concentration of births, weak productivity, and by a short production cycle. This has a direct impact on the marketing of sheep, which is influenced by five factors:

1. The seasonal fluctuation of sheep supply is bound to climatic conditions. The livestock producers tend to guard the maximum number of animals during the period when graze is abundant. Conversely, when the pastures are exhausted pastoralists tend to sell sheep in larger numbers. The sales period extends itself from April to September. The prices evolve in an inverted sense: in winter they reach a maximum, in summer a minimum.

2. Prices for sheep meat are fixed by the Moroccan government at the wholesale and retail level and are rather stable, while prices for live animals are not fixed and subject to negotiations.

3. The sacrifice festival (Aid el Adha) – an occasion on which most families slaughter a sheep – has a high impact on commercialization. Its date changes according to the Muslim calendar and rotates slowly during the year, independently of agricultural seasons. In the weeks preceding the festival, demand increases considerably, resulting in a very strong price increase for live animals.

4. The absence of a refrigeration network limits storage possibilities.

5. Prices vary according to many factors, such as: climatic conditions, race, breed, age, fodder supply and condition of the animals.

Within the marketing chain that links rural production with (urban) consumption, two groups of actors have to be distinguished. The *maquignons* are rural or urban based traders who buy, transport, and sell live animals, but they are not allowed to slaughter. The *chevillards* on the other hand are urban based wholesalers who buy livestock, slaughter it, and sell the carcass to butchers. Normally, sheep are sold to the cities via the intermediation of one or several *maquignons*, and finally end up in the hands of a *chevillard.* In general, as consumption within the eastern Moroccan highlands is, as noted, almost insignificant, urban demand shapes the provisioning chain (Fig. 7-1). Together, the four most important urban destinations receive 52% of all sheep passing through the study zone.[3] Most of this trade, once arrived in the city, is handled by *chevillards*: Of the overall *chevillard* trade originating from the study zone, 35% goes to Casablanca, 16% to Oujda, 15% to Tanger, and 12% to Rabat; the remainder goes elsewhere. From the perspective of the cities the relative importance of this supply chain is as follows: Casablanca receives 31% of its sheep meat from the study area, Oujda 65%, and Rabat 25%. Northeastern Morocco and its urbanized areas – mainly the city of Oujda, and the regions of Nador and Berkane – constitute a third of the overall demand.

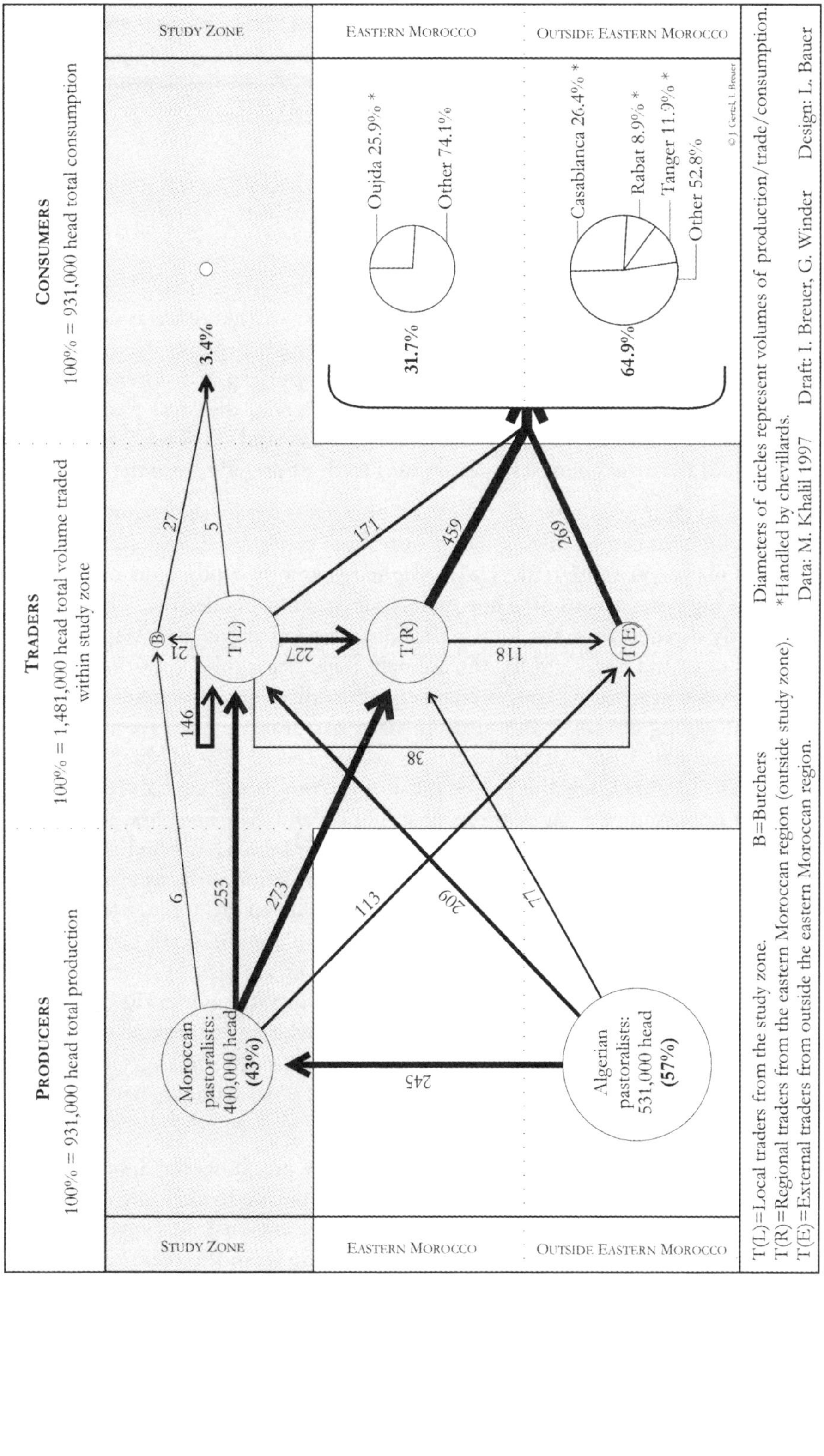

Figure 7-1: Sheep Commodity Chains in Eastern Morocco (Numbers in 1,000 Head)

In the following, two case studies will be discussed. The first one will provide a more detailed analysis of the rural-urban commodity chain, specifying the role of the *maquignons.* The second case study will focus on the livestock producers themselves revealing the role they play in different marketing chains.

The *Maquignons* and the Rural-Urban Commodity Chains

The study area is characterized by a very intense livestock trade. Figure 7-1 demonstrates the circuits of ovine commercialization in the zone, as well as their linkages with external markets. The core of the system depends on local pastoral production comprising about 400,000 head. By importing and fattening an additional 245,000 head, probably originating from Algeria, the local pastoralists sell 645,000 animals annually. 59% of these animals are sold on weekly markets and 33% are sold to *maquignons* who are coming to draft directly from the camp sites.

According to their positions, strategies and objectives within the commodity chain, these *maquignons* can be distinguished into three types: local, regional, and external. Local *maquignons* are traders who originate from the study zone in the eastern Moroccan high plateaus itself. They mainly act as animal collectors: They buy animals mostly directly from the local pastoralists, roving about the far-spread campsites to choose and negotiate for the animals. This situation allows them to profit from the weak negotiating capacity and the immediate needs of money, especially prominent among the small and medium sized pastoralists. They are also involved in buying animals from Algeria, which make up about 45% of their commercial activity. The local *maquignons* re-sell most of the purchased animals on the zone's two most important weekly markets at Tendrara and Ain Beni Mathar, but they also export a substantial part of the volume of their activity beyond the high plateau. Their main clients are the regional *maquignons* from northeastern Morocco – especially those from the town of Taourirt – followed by traders from Oujda, Casablanca, Rabat and Tanger. They also provision for the small circuit of local consumption. The majority of the local *maquignons* are also by themselves livestock producers. Some of them live in town, others in a tent out in the open steppe. Many of the local *maquignons* act as small traders, who enter the *souk* in the morning without money and leave a few hours later with a small profit varying from day to day. Only very few among them are so-called *sbaybias* – traders having financial means and being regarded as more 'professional'.

The most important actors of the sheep commerce are, however, traders who do not originate from the study zone itself. They can be subdivided into two groups, namely: the regional and external *maquignons*. The regional *maquignons* – at the time of the study, their number was 121 – originate from the regions adjacent to the study area in northeastern Morocco. The majority of them (71 of 121) is from the town of Taourirt. Unlike the local *maquignons*, 94% of them work as full-time traders and are professionals. Nearly 40% have permanent employees. Generally, they are much less implicated in trans-border trade than the local *maquignons*. They purchase livestock rather from livestock producers or *maquignons* of the study zone, mainly on the weekly markets, but much less on the camp sites. To the camp sites they go mainly for special occasions, such as national or Muslim festivals when heavy consumption arises in the urban centers, or during spring time when supply from the pastoralists is abundant. They customarily buy young male-goats or lambs. The regional *maquignons* are marketing about 42% of the animals sold by the livestock producers. To do so they develop very strong relations with the local pastoralists. If demanded by a producer, they may buy animals even when they do not have the intention of buying. The same is true vice versa: they may demand of a livestock breeder to sell some of his animals at some point in time, even if he does not intend to sell any. On the other hand the regional *maquignons* entertain strong links with urban *chevillards* with whom they may cooperate over a long time. Not less than two thirds of their stock turnover go to far-away regions of the country. They travel a lot for purchases and sales, often up to 800–1,000 kilometers. The regional *maquignons* may sell on the market of Taourirt where they meet *chevillards* from far-away regions, but they may also transport the animals to sell them on *souks* of other regions in the country, notably in Ben Guerir, Casablanca, Khouribga, Fqih Ben Salah, Beni Mellal, and Sidi Hajjaj. However, their main clients are *chevillards* from the large Moroccan cities, mainly from Casablanca, Rabat, Oujda, and Tanger, but also from El Hoceima, Agadir and Marrakech. Extended social networks thus constitute the most fundamental determinant for success in this group of actors. Those who succeed, entertain good relations with external actors, notably those in the cities. Telephone communication plays a key role in sales by confidence and by command, but this kind of trade is only realized with the *chevillards* from Casablanca. An important marketing strategy of the regional *maquignons* is to fatten the animals with lucerne, bran and barley. 80% of them employ feedlots. They then sell the animals during periods of high demand, above all in winter and during festivals. In this way, they contribute to a temporal market regulation. The regional *maquignons* are also drafting their animals in order to constitute homogeneous flocks for their respective customers, since pastoralists sell without this kind of selection. Closely interrelated with both the top and the bottom of the commodity chain, the regional *maquignons* constitute a key group in the sheep marketing system of the eastern Moroccan high plateaus.

The external *maquignons* – 44 persons – originate from far away regions, such as Marrakech, Oued Zem, Khouribga, Beni Mellal, Er Rachidia, Ouarzazate, Chichaoua, Settat, Laayoune, or Tanger. They commercialize 30% of the entire volume issuing from the zone. Only 50% of them work for their own pocket. Very often, they are engaged as intermediaries for distant actors. 64% of them work full-time. Their mean annual trading volume is much lower than that of a regional *maquignon* and only 20% of them fatten their animals. External *maquignons* solely purchase on the *souks* and not at the camp site, but they play an important role in supplying the region with animal fodder, primarily barley. In return, they sell live animals to urban *chevillards* and also to retail butchers or consumers of their regions of origin, particularly during festivals. As different consumption habits are shaping local taste, for example as consumers prefer meat of a certain breed or age of animal, these *maquignons* are also specializing to trade certain types of animals.

Pastoralists and the *Oulad Jellal* Commodity Chain

The pastoralists of the study area are part of the Beni Guil tribal confederation. Their livelihoods have been the object of several studies (cf. Rachik, Mahdi, this vol.), but the importance of their livestock marketing strategies have rarely been studied, which will be undertaken in the following, in relation to different types of pastoralists.

Traditionally, the zone's pastoralists breed sheep of the local *Beni Guil* breed. At present, the entries of Algerian sheep foster an unstable supply structure and fluctuating prices both depending largely on the vigor of border controls. But the 'imported' sheep is also of a different breed called *Oulad Jellal*, the most important Algerian breed. Its net reproductive performances are better in comparison to the *Beni Guil* and certain urban *chevillards* prefer this race also because of its butcher-related qualities. The *Beni Guil* sheep on the other hand is on high demand at the *Aid el Adha* celebration. It is the presence of these two breeds on the local markets that assigns a considerable dynamism to the local trading activities that pastoralists are involved in.

In the study area, the system of sheep raising is marked by uneven herd repartition between pastoralists. The basic herd sizes vary between 15 and 4,500 ewes. Large herds can be observed in the southern part of the study area, while herds of high quality are frequently found in the northern part. According to herd size and commercial practices four groups of pastoralists can be distinguished.

First, pastoralists with small sized herds: Pastoralists with the smallest herds possess an average of 56 ewes. 70% of this group originate from the northern part of the study area. Their reproductive herds are marked by a weak presence of the *Oulad Jellal* breed – namely only 8%. This can be explained by the limited financial resources among this group, since the raising of the Algerian race requires much supplementary feed. Nevertheless, these pastoralists sell much more animals than they produce annually – 131% in comparison to their basic herd, revealing their engagement in sheep speculation. They not only sell their own animals, but they also purchase from other livestock producers. Concerning their sales, we can say that their principal customers are local (41%) and regional (53%) *maquignons*.

Second, pastoralists with middle sized herds: Their average herd size comprises 155 ewes, 47% of which are from the *Oulad Jellal* breed. All of these pastoralists originate from the northern part of the study area. Their most apparent characteristic is a sales volume of about 152% in relation to their basic herd size, revealing that they are strongly articulated with the transborder *Oulad Jellal* commerce. They mainly sell to the regional *maquignons* who constitute 87% of their clientele.

Third, pastoralists with large sized herds: These pastoralists own on average 266 ewes. They predominantly originate from the southern part of the zone (80%). Of the four groups, they have the highest percentage of *Oulad Jellal* sheep in their herds – namely 52%. Their sales volume is 100% in relation to their herd size. This indicates that they purchase animals, but at a much smaller scale than the first and second group. The most outstanding characteristic of this group is its mode of marketing. They seem to be specialized on selling directly to the *chevillards* of the city of Oujda, who strongly demand the *Oulad Jellal* for butchering.

Fourth, pastoralists with very large sized herds: These pastoralists own on average 1,105 ewes. The majority of them (85%) originates from the south. 42% of their herds are *Oulad Jellal.* This group certainly has the necessary financial means to keep this breed. They provide their animals with lots of purchased fodder. The group sells 57% in relation to its basic herd. This comparatively weak percentage indicates that they are not engaged in speculation, which may partially be due to the fact that, being full-time livestock producers, they lack the time. They sell 74% of their sheep production to regional *maquignons* and do not have any direct business ties with the *chevillards.*

The commercial practices vary from one group to the other in the volumes of purchase and sale, and also according to the type of customers. The comparatively highest sales volumes are accomplished by the middle sized pastoralists, followed by the small. The same is true for purchases. Hence, the two smallest groups are most intensively engaging in speculative sheep trade, which is in turn articulated with cross-border movements. The small livestock producers also present the lowest rate of penetration by the *Oulad Jellal* breed, but are still very much engaged in sheep traffic. Thus, even if they live in the same place and use the same pastures, different pastoral groups may, according to their respective possibilities, engage in different marketing chains. Moreover, the presence of the *Oulad Jellal* race on the market and the cross-border price differences create an opportunity for speculation to the small and medium livestock producers. It allows them to introduce quantities of animals into the market which are much higher than their productive capacities.

Conclusion

The region of the eastern Moroccan high plateaus is a zone with a very high potential for sheep production. At the same time, the region experiences the occasional massive entry of animals of the *Oulad Jellal* breed coming from Algeria, fostering a commercial dynamism in the region. The plateaus are thus not only the principal supplier for large Moroccan cities, but they also provide other regions with sheep meat, particularly in winter. Hence, the local market remains very dependent on the large and distant consumption centers. Subsequently, numerous actors intervene at different levels of the commodity chain, most importantly, the *maquignons* with adjusted strategies for different market segments. From the perspective of the urban *chevillards* the live sheep market on the high plateaus is interesting, especially due to low prices and the presence of the *Oulad Jellal* breed. At the production level, again specialization takes place; it is the small livestock producers who engage in livestock speculation, linked to the cross-border movements of *Oulad Jellal* sheep. The sheep commodity chain and the system of commercialization thus comprise different economic circuits linking rural livestock production and a predominantly urban demand. Generally, the *maquignons* can be differentiated according to the marketing strategies they employ: The local *maquignons* play the role of collectors and small-scale traders. Next to the pastoralists they are the principal actors to supply the markets with animals originating from Algeria. The regional *maquignons* assure the zone's linkage with other regions of the country. They perform necessary adjustments, like the separation of herds, fattening and reallotment. Consequently they also assure a regulation of the market. They entertain good relations with external actors and preeminently the *chevillards*. The external *maquignons* assume the role of intermediaries for even more distant *maquignons*. They often serve very distant markets and sometimes specialize on specific market niches. Besides, they also supply the zone with fodder, above all barley. Simultaneously, the system of commercialization remains strongly articulated with pastoralism. Many *maquignons* are livestock raisers themselves. Trade and livestock production are thus extremely entrenched activities.

Notes

(1) The following text is based on Mohammed Khalil's (1997) unpublished study "La filière de la viande ovine sur les Hauts Plateaux de l'Oriental au Maroc." The text was compiled by Ingo Breuer and translated by Sandra Calkins, and summarizes only a part of Khalil's findings and argument.

(2) Projet de Développement Pastoral et de l'Elevage de l'Oriental.

(3) For methodological reasons, the destination of only a part of the sheep leaving the zone could be determined – those handled by the *chevillards*. It is very probable that a certain amount of the sheep consumption specified as 'other regional' and 'extra-regional' in Fig. 7-1 is also consumed in these four cities. Their share of total consumption would then be higher than indicated here.

Marketing from the Margins: The Ilimchan Pastoralists of the Pre-Sahara

Ingo Breuer

The present paper examines structures and dynamics of livestock marketing systems in remote pastoral areas, exemplified in a case study from the Jebel Saghro in the Moroccan Pre-Sahara. It does so from a local producer perspective, addressing livestock producers of a region that is commonly perceived as spatially, socially, and economically peripheral and remote.[1] This image is certainly not false since a significant part of the Saghro's population is subject to harsh living conditions. However, the local society is extremely polarized, and while some of the pastoralists lack access to the most basic resources for survival, others have managed to achieve livelihood security and build up wealth. Pastoralists in the Saghro also show very different degrees of involvement in various livestock marketing options, some of which are more profitable, others less so. Participation in rural-urban livestock commodity chains may vary greatly among the producers of even the same remote village, different pastoralists selling, for example, in different marketplaces or at different times. Concomitantly, pastoralists may display varying capabilities to make use of and benefit from market structures and dynamics such as price fluctuations. The paper departs from the assumption that there are multiple and close interrelations between pastoralists' market integration, their command over livelihood assets,[2] and the degree of livelihood security they may achieve. It shall be discussed under which conditions livestock producers are actively involved in – or are excluded from – rural-urban livestock commodity chains. Pastoralists' marketing capabilities are, as it will be shown, deeply rooted in resource endowment and access: not everybody can just 'sell where he likes'; on the contrary, many of the marketing strategies, especially the more profitable ones, demand considerable material and financial means, for example when it becomes necessary to constantly maintain high degrees of spatial mobility. In this context, the paper aims at exploring, for the Jebel Saghro case, the nexus between pastoral livelihoods, and structures and dynamics of rural-urban livestock commodity chains.

On a conceptual level, the paper's argumentation refers to 'real market' approaches in the sense of Harriss-White (1999), Mackintosh (1990), and Alcantara (1993). These emphasize the (local) specificity of actually existing markets and the embeddedness of exchange processes in economic, political, and social structures and institutions. The paper also implicitly draws on concepts from the French *filière* approaches, which focus on commodity-specific dynamics of exchange in studying the respective processes and transactions from primary production to consumption (Raikes et al. 2000).[3] The paper will begin with a top-down analysis of the socio-economic, geographic and institutional structures of the Moroccan sheep-marketing system, in order to sketch the framework within which the Saghro's pastoralists market their livestock.[4] After that, a local producer-centered perspective will be taken, focusing on the Ilimchan pastoralists of Iknioun in the Jebel Saghro, a pastoral community located at the 'margins of Morocco'. After an account of the local socio-economic setting, the livestock producers' situation will be addressed by a discussion of the main axes of social stratification among them. Finally, the paper will proceed to an analysis of their marketing strategies. All of the findings concerning the Ilimchan originate from my fieldwork undertaken in 2002/3.[5] Structures of livestock markets – and the corresponding marketing patterns – are, however, subject to considerable intra- and inter-annual variation, and the time spent in the field was certainly too short to evaluate these. Hence, the scope of the findings is limited, reflecting one apparently static segment of a dynamic system. The case study reveals, however, how livestock producers in a particularly remote and peripheral place are integrated into national livestock markets, and how structures of rural-urban commodity chains may represent potentials and constraints for their livelihoods.

Structures and Dynamics of Moroccan Sheep Commodity Chains

The Moroccan sheep meat sector is characterized by significant spatial and temporal fluctuations of supply and demand. On the supply side, the large majority of sheep originate from extensive pastoral systems,[6] where production depends on factors such as the animals' reproductive cycles, and regionally specific seasonal variations of precipitation. Meat demand is mainly concentrated in the urban areas. The way of an animal from the (rural) producer, the pastoralist, to the (urban) consumer is organized in a multitude of marketing chains, many of which serve specific regional or temporal demand structures. Some of these marketing chains have been subject to research projects. Paulus et al. (1994) have published an influential study, based on case studies from the Middle Atlas. Following up, several empirical investigations on livestock markets have been realized during the 1990s, many of them originating from the two Moroccan agrarian colleges.[7] These works are of varying methodological and empirical quality. Most of them are limited to small-scale local case studies, and only few of them have been published. Khalil (1997) has put forward an analysis of outstanding empirical depth on the eastern Moroccan highlands. Despite these activities, knowledge about many of the aspects of Morocco's sheep marketing system remains scarce.

For centuries, weekly markets have been the interfaces in the traffic of goods in rural Morocco. Roughly 850 markets, held on one (or sometimes two) set days of the week, are spread over the country. Almost every one of them disposes of a livestock section.[8] The individual markets differ greatly from each other in respect to many factors, such as size, the traded animal types, the suppliers and customers encountering each other. On these markets, animals are sold alive by pastoralists to traders – the *maquignons* – and on their way to the cities they are normally re-sold one or several times between different traders before ending up in the hands of urban-based wholesalers (chevillards) who slaughter them.[9] In other cases, *maquignons* also sell live animals directly to consumers. It is the *maquignons* who exploit the temporal and spatial price fluctuations throughout the country, especially at the time of national and Islamic festivals.

Morocco's weekly livestock markets are characterized by a clear spatial hierarchy, where markets of different sizes assume different functions in the commodity chains. Several attempts have been made to develop typologies of these markets. Most often, three types are distinguished, namely: (1) small rural livestock markets, located in remote community centers; (2) middle scale collector markets, located in regional towns; and (3) middle- and large-scale distributor markets. In many regions, as most studies indicate, the smallest markets play a subordinate role for the interregional animal trade, rather functioning as food-provisioning markets for the rural population. The middle-scale markets, on the contrary, have developed into the central interfaces between the pastoralists and the interregional traders, and this even more so, as they are often located in the regional towns, many of which have recently known extraordinary population growth.

The Moroccan sheep meat sector is so far almost completely sealed off from the world market. Albeit officially foreign trade liberalization has taken place, de facto high tariffs still protect domestic producers from foreign competition. There have been international agreements on gradual dismantling of these tariffs but it is difficult to predict when and how this will be realized. There are no official imports or exports of sheep meat or live sheep of a considerable volume, but there is remarkable smuggling of sheep between Morocco and Algeria (cf. Khalil, this vol.). The sheep produced by the country's pastoralists are almost exclusively consumed domestically. The pastoralists' marketing opportunities are therefore closely linked to recent developments in Morocco's socio-spatial structure. There are several dimensions to this. For one, there is a deep urban bias in meat consumption, which is heavily concentrated in the cities, whereas the rural population consumes only very little meat (cf. Chiche, this vol.).[10] In the context of increasing urbanization, provisioning the urban population with meat from rural areas becomes more and more important. Second, the annual per capita meat consumption in Morocco is still very low – 17.2 kg in 1999 –, and it has been almost at the same level for more than three decades. This seems to be connected to the stagnating economic condition of large parts of the population. For the poor, both urban and rural – but especially in the countryside – meat remains a luxury item that is consumed rarely and in small quantities. At the same time, Morocco has seen the emergence of a predominantly urban-based middle class with a higher meat demand.[11] Most importantly, however, meat consumption patterns have changed. The share of sheep and other red meats in total meat consumption has dramatically dropped within the last three decades, from 88% of meat weight in 1970 to 52% in 1999. This decline is related to the state-subsidized development of a modern poultry sector, and due to the fact that poultry consumer prices are considerably lower than those of red meat. Despite the decline of its relative importance, the absolute volume of sheep meat consumption has slightly increased during the last decades as a result of population growth.[12]

State interventions in the meat markets occur on two levels: First, rural-urban and interurban meat transports are forbidden in Morocco, due to the limited availability of refrigerated transport equipment. In the entire meat-marketing system, the animals are thus transported alive and are generally slaughtered in the place of consumption. Second, the prices for most types of meat are fixed on the wholesale and retail levels. Price fixations are arranged by provincial or municipal authorities, and prices may vary slightly from one city to the other. Fixed prices generally are only valid for meat sold according to the so-called *coupe traditionelle*, where the carcass is cut apart without distinction between choice parts. This practice accounts for 80% to 90% of sheep meat sales in the cities, and nearly 100% in rural areas (Paulus et al. 1994, 100f.). In contrast to this, in the *coupe européenne*, where the carcass is divided with a clear distinction between steak, shoulder, etc., the final sales prices are free.[13] Furthermore, while consumer prices for meat are fixed, there are no price regulations on the level of live animal trade. Most importantly, this means that the entire *Aid el Adha* trade, in which prices may be very high, is not subject to price regulation. Thus, in the Moroccan marketing system that is characterized by heavy spatial and temporal fluctuations in supply and demand, pastoralists and animal merchants can negotiate prices.

There is a strong cultural and religious dimension to Moroccan sheep meat consumption patterns. Especially the male sheep is, as Chiche puts it, "a prestige product, sold at high prices, ... reserved for special occasions" (2001, 266; my translation from French). In fact, the relevance of religious festivals for the entire meat-marketing system can hardly be overestimated. The event most overwhelmingly influencing sheep demand is the Islamic *Aid el Adha* festival, which occasions the sacrifice of one sheep by each family head. The number of sheep slaughtered in Morocco each *Aid el Adha* is estimated at 3.5 to 4 million, or roughly 50% of all sheep slaughtered in an entire year throughout the country. Half of the demand is thus concentrated in one very short period of the year.[14] Dates of *Aid el Adha* abide by the Islamic calendar, and they rotate according to the Gregorian calendar. For example, in 1998 it was celebrated in April, in the year 2005 in January. Its temporal position in relation to the animal production cycles in the different Moroccan regions differs thus from year to year. For many pastoralists, *Aid el Adha* is the single most important marketing season.

Apart from its role in the festivals, it is very difficult to evaluate the position of sheep meat in urban meat consumption. In Morocco, consumer preferences for certain kinds of meat are based on a great variety of cultural norms, which may differ vastly according to an individual's region of origin or his socialization (cf. Chiche 2001). Likewise, different sheep breeds may receive variable ratings for taste and quality from distinct consumer groups. As Khalil (1997) points out, animals of different race, size, origin and age may be preferred in certain cities and at certain festivals, and throughout the country, there are thus numerous local niches for the marketing of sheep of defined age, size, race, and taste. The *Rahhali* breed of the Ilimchan does – opposed to several other Moroccan breeds (cf. Khalil, this vol.) – not have any special reputation for its taste, and it is subject to strong competition from other races which are more appreciated. There are, however, indications that there is a market niche for *Rahhali* sheep at *Aid el Adha*: Whereas, by many urban families, large sized male sheep of other races are preferred for sacrifice, smaller-sized sheep of breeds such as the *Rahhali* seem to be in demand among the urban poor who cannot afford large-sized and well-fattened sheep of the other breeds.

The Moroccan sheep meat markets offer, summed up, considerable marketing possibilities for extensive mobile livestock producers such as the Ilimchan pastoralists. Population increase, domestic market protection, fixed prices for meat, and *Aid el Adha*, are factors which thus far have assured a meat demand that is, in the long run, relatively stable. In the short run, however, the Moroccan sheep meat sector is subject to remarkable temporal fluctuations of supply and demand. Benefits from livestock marketing depend on the exploitation of price margins between different places or seasons, especially during Islamic festivals.

Jebel Saghro and the Ilimchan Pastoralists

The Jebel Saghro, an arid mountain range characterized by high plateaus and rugged gorges, is situated in southeastern Morocco, between the High Atlas and the Pre-Saharan plains, at an altitude of between 1,000 and 2,700 m. The rural commune of Iknioun, belonging to the province of Ouarzazate, covers large parts of the central Jebel Saghro; in 2004, its population was estimated at 15,738 persons (1,645 households), scattered in about 45 villages, many of them segmented into numerous isolated hamlets. Iknioun, the community center, counts a permanent population of about 450, and has a basic infrastructure with a weekly market, communal administration, schools, a ward, and some modest shops. It is linked to the next town, Boumalne, by a 45 km gravel road that was scheduled to be asphalted in 2004. The villages and hamlets, situated at up to 40 km from Iknioun, are accessible by dirt roads of changing quality, and some of the more remote locations can be reached by mule only. Iknioun's population consists of several fractions pertaining to the former tribal confederation of the Ait Atta. The Ilimchan are the largest tribal fraction of the community. Livelihoods in the area used to be entirely based on mobile livestock production till the 1930s, and until that time, virtually no permanent dwellings existed in the central Saghro, as the population lived in tents. Since then, however, the nomads have embarked on a slow and gradual process of sedentarization, which was nearly accomplished by the late 1990s, when all of them possessed fixed houses.

Mobile livestock production has declined in the last few decades. According to official data from the local agricultural administration, there are an estimated 14,000–25,000 sheep, 15,000 goats, and 300 dromedaries in Iknioun.[15] The percentage of households still active in mobile herding varies from village to village; it is around 5–10% in most cases, but may reach up to 30% in some. The majority of households, however, only keeps small herds of sheep and goats (mostly 30 head and less), which are pastured around the villages and put into stables at night. Today (i.e. in 2002), most households depend on varying combinations of occasional labor income, livestock production, the gathering of plants and herbs, and subsistence-oriented agriculture. Some locals have been working as soldiers in the Moroccan army throughout the country, leaving their wives and children at home. This seems to be one of the few stable income sources in the region. Many younger men are active as day-laborers in the neighboring oases and in northern Moroccan agricultural areas and cities, mainly Rabat and Casablanca, but during the last years income has been very infrequent in most cases, as many did not manage to find work for more than some very short periods.

For several decades now, the Ilimchan have engaged in agricultural production. Around each village, barley, maize, wheat, and lucerne are cultivated. Outputs are entirely destined for household consumption or used to feed the domestic animals. Agriculture has always been limited by water scarcity, especially in the past two decades, which were characterized by numerous drought years. These years have brought about a massive expansion of private wells equipped with motorized pumps, on which most of the agriculture now depends.[16] As a consequence of the intensive exploitation of water resources, the ground water table has lowered considerably in many places, and numerous wells had dried up in 2002, considerably reducing grain harvests. Another source of livelihood is the gathering of herbs and ligneous vegetation, mainly performed in winter. These plants are used for heating, cooking, medicinal treatment, or to feed the domestic animals. In some cases, a part of this material is also commercialized. In general, subsistence-oriented activities make up substantial parts in many households' economies; especially the poor strongly depend on it. To a large extent, these activities are carried out by women, who also perform most of the everyday household work required to sustain the functioning of the more 'visible' livelihood activities such as salaried work and livestock marketing.[17]

Monetary poverty is widespread in Iknioun. In many households, cash income is subject to extreme fluctuations and a high degree of insecurity. Often, it is so low and infrequent that it is hardly possible to evaluate. Illiteracy rates are very high (83% in 1994).[18] In the National Map of Communal Poverty published in 2005, Iknioun was classified among the poorest communities of the country with a poverty rate of 49.9% (HCP 2005). In 2002, after several dry years agriculture in Iknioun was in a deplorable state. Many wells had dried up, and at least 70% of the households had hardly harvested any grains. In many cases, cash expenses had been reduced to a minimum, and people were trying to cope by intensifying their gathering activities in order to minimize cash expenses for animal fodder. The poorest of the poor seemed to live off a few goats and the odd exchange of gathered plants for a little bit of food.

But this account of livelihood deprivation represents only one side of local society, which displays extreme social differentiation. It is hard to say whether differences in resource access have been aggravated recently, but some findings indicate that processes of considerable asset concentration have occurred, most visible in agriculture and in the livestock sector. For example, many pastoralists with smaller herds who could not afford to buy supplementary fodder had been compelled to sell their herds during the recent drought years, and many of the herds in Iknioun are either very small or comparatively large (cf. Table 8-1). In agriculture, drought has unleashed a fierce water competition in some places of the community, as some families, mainly wealthier ones, dig new, ever deeper boreholes, in order to pursue market-oriented fruit and vegetable production.[19]

The following analysis of social differences among the Ilimchans' livestock producers is based on the case study of Miqbi, one of the community's largest villages, located at about 4 km northeast of the community center in Iknioun. Miqbi counts 85 households, the majority (81.2%) of which keeps herds composed of sheep and goats. The livestock-producing households can be subdivided into three groups. Table 8-1 displays, for each of these groups, their relative importance in Miqbi, as well as case studies exemplifying income sources and herd-keeping patterns to be found among the respective group.

The immobile livestock producers comprise 53.0% of all Miqbi households. They do not possess tents; animals are generally cared for by women and children. Herds are kept on pastures located in the vicinity of the village; at night, they are taken to stables or paddocks next to the houses. Animals normally receive supplement feed during the winter months, but in dry years the pastoralists may decide to extend this period. About two thirds of this group own small herds of 5–30 head only, and they depend largely on non-pastoral income sources. Among this group a few well-off households can be identified – they are, for example, active in trade or in the military – but the vast majority's economic situation is extremely precarious. The remaining third of the immobile herders possess middle-sized herds of about 50–130 head. Many in this group used to follow mobile herd keeping systems until

Table 8-1: Social Stratification in Miqbi Village According to Livestock Production

	Social group	Percentage of households in Miqbi [n=85]	Five case studies: Name of household head[#]	Demographic structure[$]	Herd size and mobility	Income sources (except agriculture and gathering) *(ranked in importance, as done by them)*	Herding tasks performed by[§]	Livestock marketing patterns *(markets, ranked in importance)*
Immobile livestock prod.	**Small-scale producers** -- herd size: 5–30 head	36.5	*Addi*	2 adult men 1 adult woman 3 children	12 (immobile)	(1) day labor in the regional oases and in Casablanca	women and children	Iknioun only
	Middle-scale producers -- herd size: 50–130 head	16.5	*Brahim*	3 adult men 3 adult women 5 children	86 (immobile)	(1) livestock sales (2) butchery (one day per week) (3) emigration*	women and children	(1) Iknioun (2) Boumalne
Mobile livestock producers	**Middle- and large-scale producers, mobile within the Jebel Saghro only** -- households split up between house and tent -- herd size: 150–400 head	7.1	*Daoud*	5 adult men 3 adult women 2 children	200 (1 tent; mobile in Saghro region)	(1) livestock sales (2) day labor in Rabat	2 adult men 2 women	(1) Boumalne and Tinghir (2) Iknioun
	Middle- and large-scale producers who relocated to the Souss in 2002 -- households split up between house and tent -- herd size: 150–800 head	21.2	*Lahcen*	5 adult men 3 adult women 5 children	600 (3 tents; mobile throughout Morocco)	(1) livestock sales	3 adult men 2 children 1 hired shepherd	(1) Souss (2) Boumalne and Tinghir (3) Iknioun
Households not producing livestock[+]		18.8	*Mohand*	1 adult man 1 adult woman 6 children	0	(1) salary from army employment	--	--

Source: Own survey. Remarks: *One son has emigrated to Spain but has not yet contributed to household income. [+]Except cattle. [#]All of the names have been changed. [$] Persons aged below 18 are considered as children. [§] Familial work force, unless classified as 'hired'.

recently, but their herd size has decreased due to factors such as poor pasture quality or the lack of family work force. Although they all dispose of non-pastoral income sources, a substantial part of their cash income still originates from animal sales.

The mobile livestock producers, in contrast, are characterized by the regular splitting up of the household between a 'nomadic' part living in one or several tents (in extreme cases up to four), and a 'sedentary' part staying in the village. The herd is accompanied by one or several adult men, their women and children, and sometimes by hired shepherds. In any case, income is based on animal sales, although in some households young men are beginning to engage in casual labor in the cities. In comparison to neighboring regions, average herd size of the mobile pastoralists is high, amounting to about 330 head. In 2002, this group was marked by a polarization of herd mobility patterns: Three quarters (18 pastoralists) had temporarily moved to the southwest Moroccan Souss region – at about 500 km from Iknioun – by truck in autumn 2002/03;[20] the remaining quarter (6 pastoralists) had moved within the Jebel Saghro region only, the distance to their home village never exceeded 50 km, and many of them kept the animals in stables during the winter months.

As the entire local society of Iknioun, the animal producers are characterized by extreme differences in wealth. Despite the preceding drought years, most of the large-scale mobile pastoralists seemed quite well off in 2002. In contrast, drought had severely hit those with small, immobile herds, and even more severely those with irregular cash income who were largely depending on subsistence-oriented agricultural production.

Livestock Marketing Structures and Strategies among the Ilimchan

Animal sales represent an important source of cash income for many of the Ilimchan livestock producers. As there are considerable seasonal fluctuations in animal prices, and because different marketplaces provide different sales opportunities, it is important for a pastoralist when and where he sells. Ilimchan pastoralists have developed varied marketing strategies aiming at adapting sales to price fluctuations. Two aspects of these strategies will be discussed in the following, one concerning marketplaces ('who sells where?') and one marketing seasons ('who sells when?'). Since every temporal and spatial adaptation of animal sales demands considerable efforts, the marketing patterns pursued by a pastoralist depend, as it will be shown, on his economic situation and asset endowment.

Who Sells Where?

In 2002, three marketplaces were frequented by Ilimchan pastoralists. First, the local weekly market of Iknioun; second, the weekly markets of Tinghir and Boumalne, located in the oases neighboring the Saghro; and third, several markets of the Souss region, around Agadir, immediately serving that agglomeration's demand. In all of these markets, both livestock and other consumer goods are traded. There are marked differences in the use of these marketplaces among the Ilimchan (cf. Table 8-1).[21] Whereas many of the small-scale producers had, during 2002, sold in Iknioun only, pastoralists with larger herds had used at least two or three external marketplaces. Furthermore, the following will show that the marketing practices adopted by the pastoralists – for example, sales organization, information acquisition, and market control – vary widely between the marketplaces.

In Iknioun, the market is held every Tuesday. In general, it mainly provides the local population with staple foods, household items, and animal fodder, and it also assumes an essentially social function. Traveling to Iknioun on a market day serves the purpose of fulfilling administrative tasks with the public authorities, or to visit the local ward in case of sickness. Also, it offers the opportunity to meet relatives, acquaintances and friends, to exchange news and information, to get involved in local politics, and to cultivate relationships. In any case, for those living in the more remote hamlets traveling to the market center involves considerable efforts, since the one-way trek generally takes several hours to half a day on foot. Though it is located in the midst of the Ilimchans' pastoral area, animal demand and supply were very weak during autumn 2002. Hardly more than 20–50 animals were offered on a typical market day, most of them goats, and many in poor condition. Sellers were exclusively local pastoralists coming to sell animals in order to receive cash for their weekly provisions with staple food and other items. The majority of the sellers only offered one or two animals each. Trader activity was sparse, and only few merchants, mainly originating from Iknioun itself, were present. Demand came mainly from local butchers, who in turn sell to the few locals who can afford to eat meat.[22] Merchants from far-away regions frequent this market in summer only, when the pastures are exhausted and pastoralists tend to sell animals in larger numbers.

The analysis of household data revealed several tendencies concerning the pastoralists' marketing practices. For example, very few of the livestock produced by the Ilimchan are sold in Iknioun. Pastoralists wanting to sell a larger number of animals at a time usually prefer the markets of Boumalne and Tinghir. Especially those with larger herds claim that it is hard to sell more than one or two animals at a time in Iknioun: "This is remote territory. It is not like in the big cities, where you find someone who will buy, no matter when you go there. There are only few traders here. You have to wait a long time for the traders to come, and after a long wait you will end up selling at a low price." Second and inversely, the poorest pastoralists seldom sell their animals at any market except the one of Iknioun. The efforts to go to one of the larger markets often seem to exceed the benefit. As one pastoralist put it: "Others tell me that in Boumalne prices are better. But look: I just want to sell one little goat and to buy something to eat from the money. It is not worth the effort to go to Boumalne just for this."

On a regional scale, the most important markets for the Ilimchan are those of Boumalne and Tinghir, two towns located in the Dades and Todgha oases north of the Saghro. At a distance of 45 and 60 km from Iknioun respectively, they have weekly markets largely exceeding in importance the one in Iknioun. If pastoralists go there, they use public transport, where animals are carried on the roof of buses or collective taxis. These are much larger livestock markets with a wide range of traders present. For example, in the first week of Ramadan in November 2002, 820 animals were offered in Boumalne and 770 in Tinghir, while the offer did not exceed 45 in Iknioun. A wide range of traders can usually be found here, and opportunities to sell large quantities of animals are much better than in Iknioun.

Several factors have influenced the dynamics of these markets. There is considerable local demand, as these oases are densely populated and have been subject to rapid urbanization in the last decades. Massive emigration to Europe has led to the emergence of groups with considerable purchasing power and higher meat consumption. In this context, a certain number of locals regularly buy animals for fattening. These markets also have regular trade links with the major Moroccan cities – assured by both local traders and traders coming from outside, especially from Agadir and Marrakech. This is partly due to the local supply structure, which offers animals of different breeds – for example, the sedentary *Dman* sheep of the oases as well as the nomadic *Rahhali* sheep of the Saghro and the High Atlas – and different ages, qualities, and fattening status. As it became evident in the household analysis, these markets are frequented by Iknioun's middle- and large-scale producers, especially when they want to sell off large quantities of animals at a time. The poor among the Ilimchan only very rarely go there. In general, the decision to take animals to Boumalne or Tinghir is preceded by a diligent quest for information about expected prices and tendencies in price development.

In the Souss, marketing conditions are even better. Within the past decades the region has experienced a strong economic dynamism, among other factors due to the growth of export-oriented agricultural production and tourism, and also due to rapid urbanization and a boom in the construction sector. As the Souss counts an urban population nearing one million, there is a constant demand of meat throughout the entire year, and especially on religious occasions demand skyrockets. Local demand is diversified and allows for several segmented niches, different consumers preferring, for instance, meat from different sheep breeds. Many of the traders who regularly come to Boumalne and Tinghir are based here, and most of them engage in fattening activities.

The 18 Ilimchan pastoralists who stayed several months in the Souss during winter 2002/03, had deliberately chosen to stay in very remote areas of the region, in order not to get involved into violent conflicts with other nomadic groups and the sedentary population (cf. Turner, this vol.). However, they were still making use of the urban marketing opportunities. For example, several of them were sojourning in a remote area in the hills of the western Anti-Atlas. They sold their sheep in Had Boulfaa, a marketplace located 50 km away from their campsites and directly next to one of the main roads of the Souss, not far from the urban fringes of Agadir, the most important agglomeration of the region. One of them relates: "There were several Ilimchan tents staying in the same region. When we wanted to sell a whole lot of animals, we charged two of us to take all of these sheep and to go to the market in behalf of all the tents. ... They stayed in the marketplace for more than one week, negotiating with many traders. Finally, they got a good price for all of us." The pastoralists also made great efforts to acquire information on price developments, a task that was not easy as reliable social networks were scarce. Some of them regularly used mobile telephones to stay in contact and to inform each other. For all of them, staying in the Souss in winter 2002/03 meant extraordinarily good marketing conditions, because this period included both Ramadan and *Aid el Adha*.

Who Sells When?

Among the Ilimchan pastoralists, three temporal patterns of livestock marketing can be identified.[23] The first one may be called the 'little-by-little pattern'. Pastoralists sell one or several animals per market day, whenever they are in need of cash for the purchase of staple foods or other everyday consumer goods. Sales are relatively evenly spread out across the year, and are also comparatively independent from seasonal price fluctuations. Equally important is a marketing pattern one may refer to as 'seasonal-en-bloc-pattern'. Here, marketing is very closely oriented towards the reproductive cycle of livestock, the conditions of production, and the changing pasture quality. Sales predominantly concern lambs born in the course of the past year, and they take place within a relatively short high season, normally during summer. The rationale behind this pattern is to keep as many animals as possible while grazing opportunities are abundant (in spring) and to sell them off as soon as graze becomes scarce. Although prices are comparatively low during summer, this is, for many pastoralists, the only season to sell off high quantities of animals, as livestock traders frequent the regional markets, which is not the case in winter. The third pattern may be called 'price-oriented livestock sale'. Here, the pastoralist adjusts his selling decision to the market conditions. He sells whenever prices rise, preferably in the periods of religious holidays, and he avoids to sell in drought seasons or periods when prices fall. This behavior allows him to skim a certain part of the opportunities for profit that is generated by seasonal price fluctuations.

Among the Ilimchan of Miqbi, different livestock producers pursue different combinations of these strategies. The poorer ones among the small- and middle-scale producers sell their animals almost exclusively according to the little-by-little pattern. Many of them dispose of only very unreliable alternative sources of cash revenue. They are thus compelled to sell animals whenever cash is needed. In low-price periods, they cannot afford to postpone sales and to wait for more favorable prices. Though estimation is difficult, this group should comprise between forty and fifty percent of the Miqbi livestock producers. Another group of the pastoralists combine little-by-little and seasonal-en-bloc-strategies, the latter frequently making up more than half of their annual livestock sales. These are generally pastoralists with middle- or large-sized herds, many of them disposing of financial resources allowing them a certain degree of flexibility in the planning of the annual household budget. Systematic adaptation of animal sales to price fluctuations is only encountered among very few of the Ilimchan pastoralists. In normal and dry years, one prerequisite to postpone sales is the capacity to keep and feed animals even in times when graze in the Saghro is insufficient.[24] Some pastoralists achieve this by temporarily relocating their herds to regions such as the Souss or the eastern Moroccan steppes. Normally, however, purely price-oriented strategies exclude relying on natural pastures alone, and the pastoralist must purchase large quantities of supplementary fodder to avoid the slimming and thus a decline in value of his stock. Temporal adaptation of animal sales thus requires considerable financial resources. Generally, only very few Ilimchan have managed to adopt this strategy. They can be found among two groups, namely the mobile pastoralists who own very large herds, and those wealthy households – with large or small herds – who dispose of profitable non-livestock income sources. In general, even these few sell only a part of their livestock according to price-oriented patterns.

Conclusion: Integration or Marginality?

The Jebel Saghro has been referred to as one of the remote and peripheral areas par-excellence of Morocco. But who of the Ilimchan pastoralists is 'on the margins', and in respect to what? From a purely territorial perspective, all of the Ilimchan are simply living on and marketing from the very margins of the country. Empirical evidence, however, showed that marginality must be conceived as manifold, in order to make sense of the multiple mechanisms of social stratification and polarization among the Ilimchan. Besides its territorial aspect, there is a social dimension of marginality, as livelihoods of the pastoral population's majority remain characterized by extreme insecurity and poverty, whereas a minority manages to concentrate considerable amounts of material and financial assets in their hands. However, marginality also has a marketing dimension. Some of the Ilimchan pastoralists are very well integrated into various, sometimes very lucrative segments of the livestock marketing chain. Others, on the contrary, are 'marginal', because they are not capable of choosing when and where to sell. They are thus exposed to spatial and temporal price fluctuations that affect their livelihoods. In any case, spatial and temporal marketing strategies are connected to and influence patterns of pastoral mobility; they are also rooted in access to assets, and thus in social differences, which they help to reproduce.

Pastoralists from remote Moroccan areas are intricately integrated into the national sheep marketing system and its numerous rural-urban exchange mechanisms. Their future livelihoods will decisively depend on the dynamics of the economical, societal and institutional framework of Moroccan meat markets. Simultaneously, however, only some will be able to participate in (new) marketing opportunities. The processes of social polarization and asset concentration that were observable among Ilimchan pastoralists during 2002 indicate that, at least in Iknioun, there might be few winners, while the majority might remain at the margins.

Notes

(1) Compare, for example, the "*Les portes du désert*"-chapter, and especially the Jebel Saghro paragraph in "*Maroc – Régions, pays, territories*" by J.-F. Troin (2002).

(2) For a detailed account on the notions 'assets' and 'resources' used here: cf. Gertel, this vol.

(3) The *filière* approach has been widely used in the analysis of Moroccan livestock markets. See Paulus et al. (1994) and the references cited below.

(4) The vast majority of the Saghro's livestock producers keep both sheep and goats. Herd composition may change among households, and also from year to year, depending among other factors on climatic variations. In 2002, in most of the studied households sheep made up between 50 and 70 percent of the herd. Yet, since sheep are more expensive than goats, they constituted a much higher part of the herds' trade value. Sheep and goats may also assume different functions within the everyday marketing patterns of households (cf., for example, Werner 2006). In the following, when herd sizes are addressed, the number of sheep and goats possessed by a pastoralist is meant. However, for simplicity, the subsequent analysis of Moroccan livestock markets will concentrate on sheep commodity chains.

(5) Between March 2002 and June 2003, the author stayed several months with the Ilimchan. Fieldwork comprised participant observation of everyday activities – including marketing – in pastoralist families, a systematic evaluation of resource endowment in 105 households, as well as regular market observations and a demographic survey of ten villages, of which Miqbi is the largest.

(6) Many of them, however, are subsequently fattened by small entrepreneurs which are

mostly based in the peri-urban areas of the large cities (cf. Chiche 2001).

(7) For example, see Khalil (1997), Dalil (2000), Benassou (1998), Rahmani (1998), Mahamat (1993). For urban sheep meat demand, see Chichaoui (2001), Marrakchi (1993). In addition, some synoptic articles about the Moroccan livestock sector have been published (Chiche 2004b, 2001; Abaab et al. 1995; Driouchi et al. 1995).

(8) The following findings are drawn from two of the most detailed studies on Moroccan livestock markets (Paulus et al. 1994, Khalil 1997). The most comprehensive typology of weekly markets in Morocco has been put forward by Troin (1974).

(9) This is a simplified and idealized account, describing the type of Moroccan sheep commodity chains that makes up for the most significant part of animal trade. Other types exist, for example rural-rural sheep marketing. Furthermore, groups of actors may overlap, as pastoralists may be active in trade (cf. Khalil, this vol.).

(10) The meat consumed by rural inhabitants also very often originates from their own production.

(11) See Chichaoui (2001) for an analysis of meat consumption patterns in households of different social groups in Casablanca.

(12) The Moroccan population has grown from 15.3 million in 1971 to roughly 30 million in 2004. The urbanization rate is estimated at about 55%.

(13) The return margin to be achieved by this mostly benefits the butchers, since these are the ones that cut the animals apart (Paulus et al. 1994).

(14) Rise in meat demand can also be observed during the Islamic fasting month of Ramadan.

(15) Data on local livestock numbers have been contradictory, and they are based on very unreliable estimations.

(16) Their number was estimated at about 1,200 in 2001.

(17) Seen from the marketing perspective adopted by this paper, women's roles in the Ilimchans' livelihoods tend to be overlooked. For example, women do not at all appear as actors in the livestock marketing system, as trade is generally the men's task. However, women perform much of the work in the livestock production process – for example, in carrying much of the herding tasks – and without their household work, the pastoral production system described here would not be possible.

(18) There have been recent state efforts to enhance access to schools, for example with the construction of a new boarding school in Iknioun. Illiteracy data were taken from the 1994 census. New data of the 2004 census were not yet available when this article was written.

(19) Most of these families had been able to mobilize capital from kin emigrated to France and Belgium in the 1960s and 1970s.

(20) For an overview and discussion of the complex situation in the Souss, cf. Turner (this vol.); Werner (2006).

(21) The data on marketing practices and marketplaces originate from household studies the author undertook during fieldwork, based on a sample of households from every of the social groups delineated above. The marketing data set included 32 households from Miqbi and 30 from other villages of Iknioun. Table 8-1 represents five selected households, each of them displaying the marketing pattern most frequently encountered in the respective social group.

(22) Mainly the small group of state employees (such as administrative personnel and teachers) living in Iknioun center.

(23) As the literature indicates, these patterns with local variations probably exist among pastoralists all over Morocco (cf. for example, Paulus et al. 1994; Khalil 1997).

(24) In years with extremely high precipitation, when pastures are in particularly good condition, flexibility in temporal sales management is higher among all of the Ilimchan except the very poor.

Animal Production, Herd Mobility, and Rangeland Access in the Middle Atlas

Houria Djoudi & Irene Hoffmann & Bouchra El Amiri & Jörg Steinbach

In the North African context, extensive sheep production systems play an important role from socio-economical, cultural and ecological points of view. Morocco's estimated national sheep herd comprises 17 million head. These animals cover more than 70% of their nutritional needs from natural pastures under various husbandry systems (Guessous 1991). However, the traditional agro-sylvo-pastoral system in the Middle Atlas mountains of Morocco has been undergoing significant changes. Due to the sedentarization tendency in the past 50 years (Pascon 1980), the privatization of land for cereal production, the restriction of collective utilization of the forest areas, and the continuously growing herd size, the pressure on the remaining collective grazing areas has been constantly rising (Guessous 1991; Bouderbala et al. 1992; Chiche 1996b; Djoudi 1998). By using communal rangeland, shepherds have developed various strategies and husbandry systems to maximize the rangeland utilization in the sheep nutrition. These strategies depend on traditional access rights to collective areas, the capital and power of the sheep owner and his ethnic affiliation (Djoudi 1998). One important aspect of the strategies in the Middle Atlas is the traditional adaptation of herd mobility to seasonal feed availability and to climatic and socio-economical conditions. The objectives of this paper are to describe the current sheep production systems, to evaluate animal performances and productivity under field conditions in each subsystem, and to determine factors that limit a sustainable use of animal and rangeland resources in the Middle Atlas.[1]

Methods and Data

The Ireklaouen area constitutes the study area for this paper. It stretches from southeast to northwest along the Middle Atlas of Morocco, at an altitude of between 600 and 2,000 m. Annual rainfall varies with altitude and ranges from 644 to 1,098 mm. Frost is common between December and March, with a maximum of 35 days with snow. Minimum temperature during the survey period (1996–1998) was -4.9°C in January. Maximum temperature was 32.8°C in July (MAMVA 1998). The Ireklaouen area covers 78,470 ha, 50% of which consists of rangelands and pastures. Sheep represent more then 89% of the total livestock while cattle and goats do not exceed four and seven percent respectively.

Based on participative research (Méthode Accélérée de Recherche Participative, MARP), 10 workshops with 10–20 shepherds were carried out in six villages and four weekly markets in 1996. Traditional property rights were discussed and maps drawn of rangelands and herd mobility, supplemented by fodder calendars. Herd management was analyzed and production constraints ranked. This participatory assessment served as a basic identification and classification of the husbandry systems in the area. Subsequently, twenty *Timahdite*-sheep herds were closely monitored between 1996 and 1998. Monthly data regarding ewe numbers, services, commercialization, and ewe live weight, lambing, abortion and deaths were recorded, as well as data regarding the number of lambs, their birth dates and birth weights, monthly live weight up to one year of age, and lamb mortality.

Additionally, the rangeland areas were identified and differentiated by vegetation type and degradation stage using the step-transect method (Norton 1995). Each vegetation type was reported at the regional map (Azrou 1/50,000). Monthly vegetation analysis was carried out 1996 to 1998 for cover, frequency, density and phytomass production using a combination of square and transect methods because of the vegetation heterogeneity (shrubs, forbs and grasses). Pastures were classified according to their botanical composition. *Genista pseudopillosa* and *Adenocarpus boudii* (pasture type I) were especially noted because of their high feed value. They are very much appreciated by herders as indicated in the MARP. Low quality pastures comprise predominately *Plantagum mauritanicum*, *Timelea virgata* and *Euphorbia sp.*, and these are classified as pasture type II (Djoudi et al. 1998). Finally, the data of the sheep flocks and range primary production were analyzed with SAS GLM and mixed procedures (SAS Version 8.2).

Traditional Conditions

The Ireklaouen belong to a major Berber ethnic group known as Beni Mguild. Originating from the High Atlas mountains, they settled in the Middle Atlas at the end of the 19th century. The Beni Mguild of the Middle Atlas include four ethnic sub-groups (tribes) with clearly delimited territories: the Ait Abdi settled in the area of the Ain Leuh community; the Ait Sgougou in the El Hammam community; the Ait Arfa du Guigou in the Timahdite community; and the Ireklaouen settled in the Azrou community area.

The map of the fodder resources, grazing areas and user groups developed during the MARP sessions indicates that twelve clans belong to the Ireklaouen (cf. Fig. 9-1). Access to range resources is handled at the clan level and not at the tribal level, and the delimitation of the rangeland areas is very strict. As in most agro-sylvo-pastoral systems in North Africa, various areas with different user rights are availa-

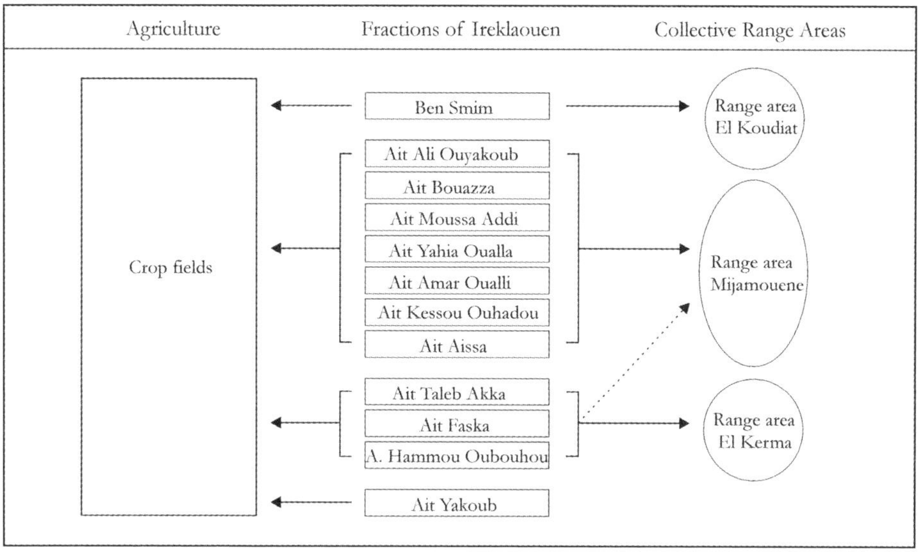

Figure 9-1: Traditional Utilization of Natural Range and Crop Areas by the Ireklaouen

ble as range resources for the sheep herds. First, collective rangeland areas in *jommouh* status: These areas are collective property, which is defined by the 1919 *dahir*. Access to such pasture is limited to defined clans. It cannot be used by other clans even of the same tribe. The most important collective pasture of the Ireklaouen, named Mijamouene (cf. Map 9-1), is located around the Hebri mountain (2,000 m MSL, mean sea level). Ain Kerma and El Koudiat, each of which has an altitude of 900 m MSL, belong as well to the Ireklaouen collective rangeland areas (cf. Fig. 9-1). Second, there are cereal cropping areas in *melk* status, which are mainly located in the valley bottoms and lowlands (600 m MSL) and have varying property regimes. They are private property during the growing cycle of the crops, but after the harvest, the stubble can be used by all other clan members as collective pasture. *Jommouh* and *melk* are governed by traditional legislation and property regimes. Third, the forest areas (Quercus ilex and Cedrus atlantica) belong to the state, and are public property (El Machzen). They add a layer of complexity to the property regimes and are a source of conflict between the shepherds and the forest administration.

In the Ireklaouen area, multifaceted and overlapping use rights to the above–mentioned areas exist (Fig. 9-1): Seven out of the twelve clans of the Ireklaouen have access to the collective rangeland area Mijamouene. These clans are Ait Ali Ouyakoub, Ait Bouazza, Ait Moussa Addi, Ait Amar Oualli, Ait Kessou Ouhadou, Ait Yahia Oualla, and Ait Aissa. Three other clans (Ait Taleb Akka, Ait Faska, Ait Hammou Oubouhou) retain the access to Ain Kerma pastures as well as to the Mijamouene area. One clan, the Ben Smim, have access rights to El Koudiat pasture.

Table 9-1: Classification of Sheep Production Systems by the Ireklaouen in the Middle Atlas

Sheep production system	Characteristics	Total time spent at the collective range areas	User groups
High-altitude nomadic	• Herd mobility within the collective range areas • Horizontal mobility (at 2000 m altitude) • No own cropping and no traditional access right to crop by-products (stubble) • No access to forest areas • Less favorable range areas • Dependent on collective water resources	12 months	Traditionally this husbandry system can be practiced by all sheepherders from the transhumant clans. At present, it is performed especially by low income shepherds. Besides the traditional right-holders flocks are also held in partnership with hired herders.
Transhumant	• Seasonal herd mobility from the cereal areas to the common range areas • Horizontal and vertical mobility • Mixed crop and animal • Possession of private cropland and traditional access to the crop by-products (stubble) • Varying levels of mechanization in animal transport, fodder supply, and cereal production • Possession of private water supply (with trucks and cisterns to the range areas) • Most herders are involved in several economic activities and political organisations	6-7 months, from March to July and/or from the beginning of October to the end of November	Ait Ali Ouyakoub Ait Bouazza Ait Moussa Addi Ait Amar Oualli Ait Kessou Ouhadou Ait Yahia Oualla Ait Aissa Ait Taleb Akka Ait Faska Ait Hammou Oubouhou Ben Smim
Lowland sedentary	• Sedentary at the private cereal areas • No access to the collective range areas • Mixed crop and animal, possession of private cereal areas and traditional access to crop by-products (stubble) • Several leasing contracts for private pastureland • Varying levels of mechanization in herd management and crop production	none	Traditionally only Ait Yakoub, but presently this group consists of high income shepherds with off-farm income, and farmers who buy land in this region and do not belong to the Ireklaouen.

This group is not entitled to use other collective grazing areas. One other clan, the Ait Yakoub do traditionally not have any direct access to collective grazing areas. However, by herding the sheep of a right-holder under traditional shareherding contracts they can also use the collective pasture.

Traditional institutions and frameworks, like the *jemaa*, are still in place for the organization of the collective use of land, water and range resources. However, due to socioeconomic and political changes, these institutions no longer perform their traditional function (Bouderbala et al. 1992; Djoudi 1996; Mahmoudi 1996), as it is the case for the *agdal*, the rotational exclusion of grazing areas for vegetation rest and recovery.

Sheep production systems in the Ireklaouen community area are thus determined by the access rights of the sheep owner's and/or shepherd's clan to natural resources, either rangeland alone, or rangeland and cropland, and the mobility needed to link these geographically separated resources. However, access to and use of collective resources depends on the status, income and power of individuals.

Three different sheep production systems were identified in the Ireklaouen area (for a detailed description see Table 9-1). In the nomadic system, herds spend the entire year in the collective range areas located in the high-altitude parts of the Ireklaouen territory. The transhumant system, in contrast, is characterized by seasonal herd mobility between the cereal areas and the high altitude pastures. In the lowland sedentary system, herds remain in the lowlands throughout the year, and no collective range areas at all are used.

Recent Transformations

Important changes to institutions and regulations for land use occurred in the colonial period, when the collective land utilization in the Middle Atlas underwent transformations similar to those operating in several other North African regions. Until then, the nomadic livestock system was characterized by large circular movements covering various ecosystems (Pascon 1980). The collective areas stood under the control of the *jemaa*. The legislation (dahir) from 27 April 1919 limited the privilege of the *jemaa* and assigned the authority of the colonial administration through the *Direction des Affaires Indigènes* (Indigenous Affairs Department) on the collective areas. With this regulation, the French administration aimed to find political and scientific answers to the question which approach to adopt with respect to the indigenous property right systems, given the goals of cost minimization, achieving land annexation for the French colonial settlers, and still endowing rural communities with a necessary subsistence minimum.[2]

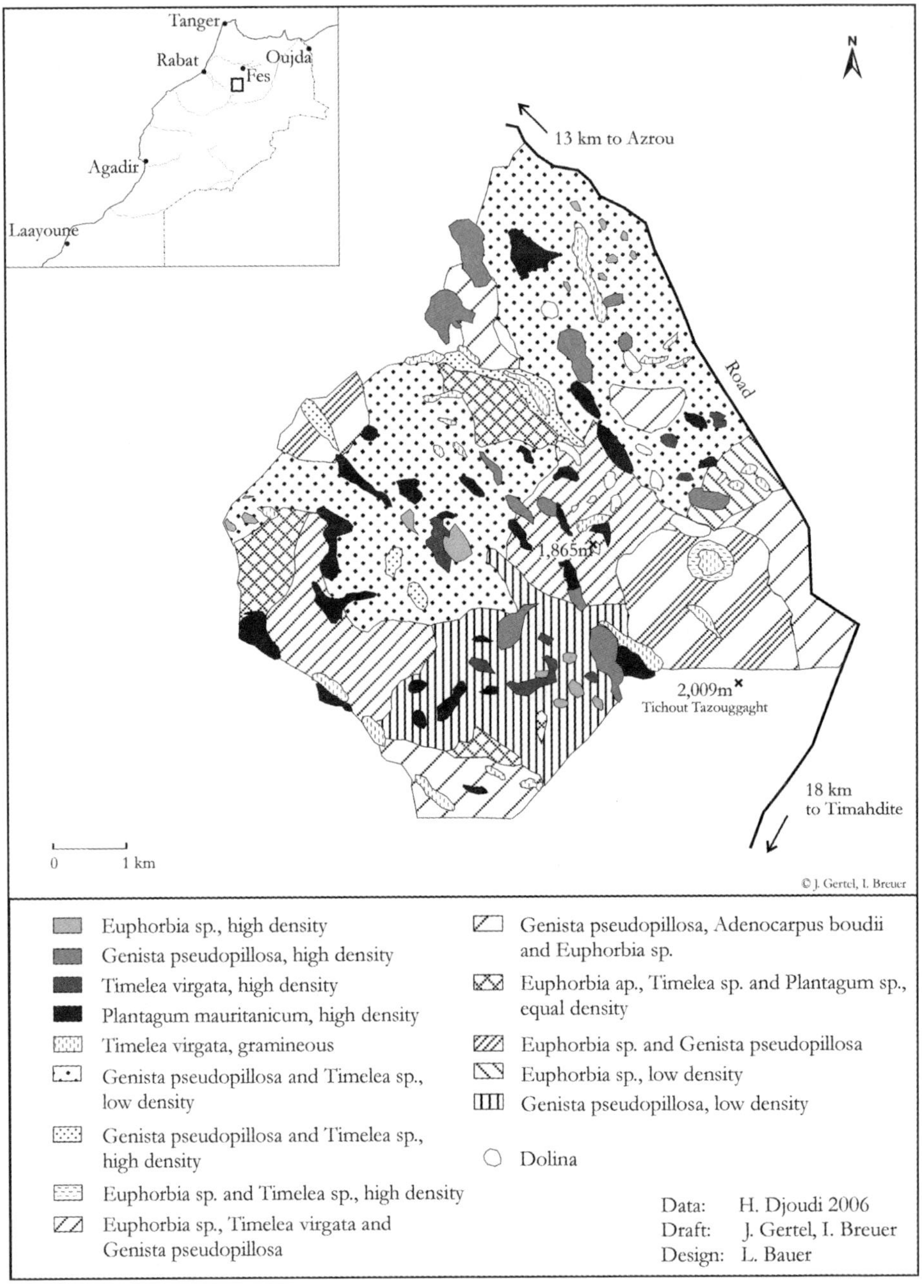

Map 9-1: Vegetation of the Collective Rangeland Area Mijamouene

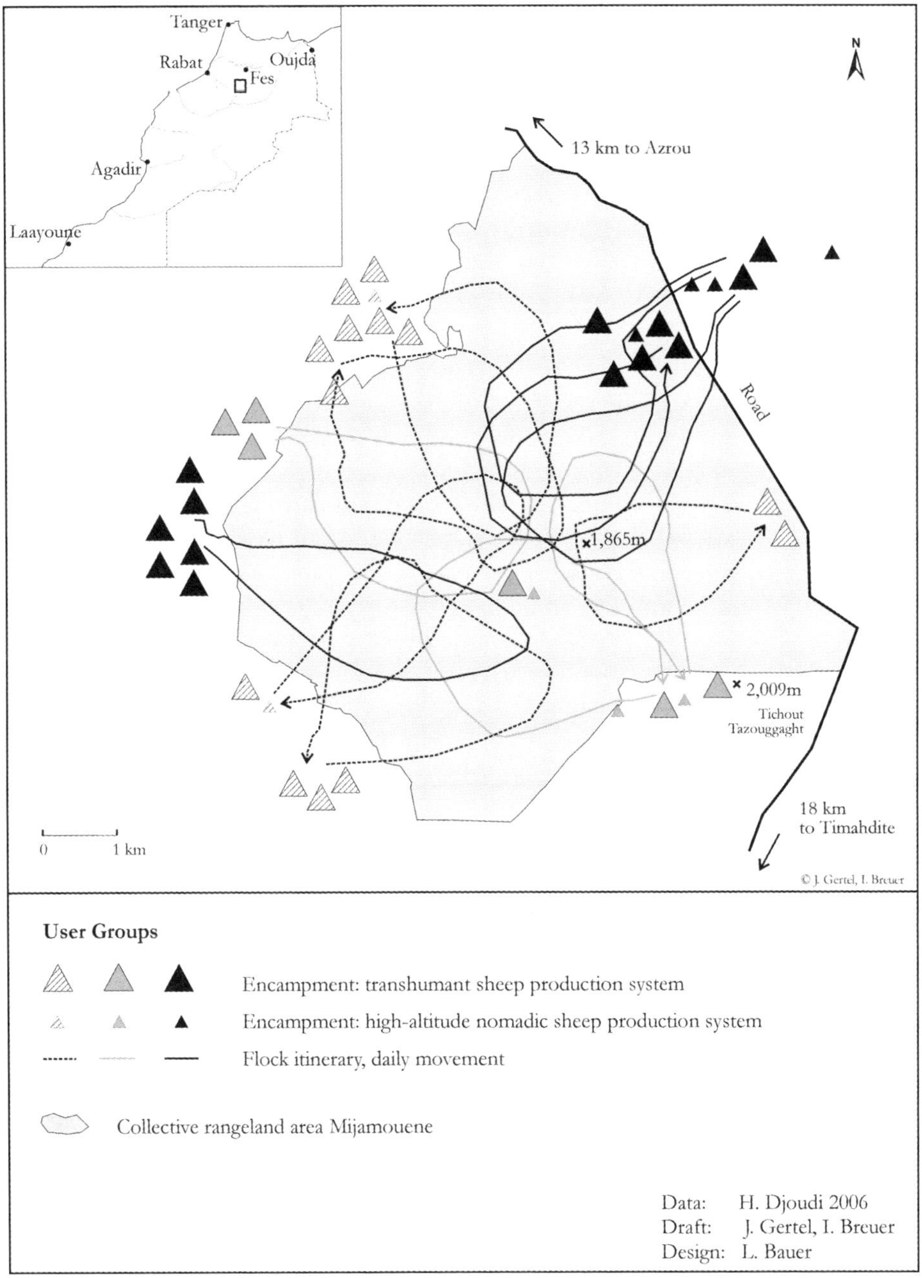

Map 9-2: Daily Movement of Sheep Flocks in the Collective Rangeland Area Mijamouene

Two years earlier, the French administration had already declared that the forest, so far a property of the tribes, belonged to the state (*dahir* of 10 October 1917). With the decrees of 1917 and 1919, the authority of the *jemaa* declined, and self-determination and autonomy of the tribes and confederations were restricted. In this same period many movement restrictions for individuals and communities were imposed, so that in several regions transhumance was controlled by the need for official authorization for passage (Laissez-passer). However the census of 1936 recorded that more than 95% of the population in the Middle Atlas were living the whole year in tents (Hsain 1989), an important indicator of frequent people and livestock movement.

Parallel to these developments, the available grazing areas in the region were reduced. The first restrictions resulted from the new forest legislation. Previously, the forests had been substantial fodder resources to assure the survival of animals during droughts. With the 1917 *dahir*, the tribes lost high-value pastures previously used during scarcity periods. The situation was further aggravated by a second factor: the privatization in the lowlands, which allowed farmers not belonging to the tribe to settle in the region. This trend became important after independence, and particularly from 1962, when farmers from other regions came in and bought the cropland of departing French farmers. This kind of privatization created a new type of farmer, possessing more land, livestock and means of production, fostering competition between individuals and decreasing solidarity among the former community members. Additionally, former institutional arrangements and regulations among different groups of users (e.g. tribes and fractions) in the area came under pressure, represented by the fact that several traditional pacts between tribes were annulled during that period, for example between the Beni Mguild and Beni Mtir (Mahmoudi 1996).

The pastoral area of the Ireklaouen borders the collective areas of the neighboring Ait Ouahi and Ait Mouli tribes. Partly as a result of the institutional changes and the reduced grazing area, the potential for conflict among the Ait Ouahi and Ait Mouli increased. The situation of those tribes was influenced by two different factors, which indirectly affected the population dynamics, the husbandry systems and increased the pressure on the remaining land of the Ireklaouen. The first factor was the gradual settlement of the Oulad Khaoua, a nomadic tribe originating from the Missour region southeast of the Middle Atlas.[3] The Oulad Khaoua are highly experienced shepherds; they immigrated into the collective pastures of the Ait Ouahi and Ait Mouli, and began to work as shepherds for them. The drought of 1947 further enforced this migration. They made use of the privatization trend in the sixties and bought large land areas in the lowlands. Despite conflicts, they continue to use the collective pastures of Ait Mouli and Ait Ouahi down to this day. The incessant conflicts about the collective areas have created a climate of insecurity and vigilance. This also affects the Ireklaouen herders: As they were fearing that Oulad Khaoua will enter Ireklaouen collective areas, several of them decided to stay there

Table 9-2: Characterization of Sedentary, Transhumant, and Nomadic Husbandry Systems in the Middle Atlas

	Lowland sedentary	Transhumant	High-altitude nomadic
Conflict risk with neighboring ethnic groups	none	low	high
Vulnerability and dependence of the collective resources (pastures and water)	none	low	high
Technical equipment for flock management (e.g. water cisterns, transport)	high	medium	low
Flock productivity	high	medium	low

all year long and give up the transhumant movement to the lowlands. The substitution of the traditional hair tents with plastic habitation in the collective areas makes it possible for herders to stay over 2,000 m MSL even in winter, although climatic conditions are very harsh.

The second factor fostering change was the installation of the Adarouch Ranch in 1966 in the collective areas belonging to Ait Ouahi and other ethnic groups. As compensation for the affected families, the state provided land near to the town of El Hajeb. The affected families, however, did not agree to leave their community and to give up other activities. They remained in the region and used traditional arrangements with Ireklaouen shepherds to keep their livestock in the collective area of this tribe.[4]

Summing up, these transformations led to an increase in the stocking rates on the collective areas and increased competition over pastoral resources. As a consequence, new forms of organization and strategies related to livestock management and husbandry systems appear in the pastoral community. New organizations are still based on the traditional ethnic framework. However, the de facto access to and the use of collective resources depend upon status, income and power of individuals rather than the traditional background. This may be illustrated by the situation on the Mijamouene pasture (Map 9-2): Everyone from the seven tribal fractions entitled to use this pasture may pitch his tent wherever he likes. Herders have different individual evaluations of the quality of certain places on the pasture; these evaluations are based on a system of indigenous knowledge called *el merged*. The map shows temporary settlements of groups of herders. Each of these groups comprises herders from several tribal fractions, which have chosen to camp near each other according to criteria such as friendship, kinship, or similar pasture evaluation (Table 9-2). During its daily movements across the Mijamouene pasture, each herd crosses zones of different ecological quality (cf. Map 9-1, Map 9-2). Herd productivity, however, depends largely, as it will be shown, on the financial background of

Table 9-3: Reproductive Performance and Productivity of Ewes According to Husbandry System and Lambing Season

Factor	Litter size	Survival rate	Parturition interval (days)	Birth weight (kg)	Weight at 120 days (kg)	Productivity index (kg/ewe/year)
Husbandry system	ns.	*	*	***	***	***
Nomadic herds	1.04 (n = 52)	0.92 (n = 152)	292 (n = 42)	3.00 (n = 382)	20.35 (n = 164)	19.78 (n = 22)
Transhumant herds	1.05 (n = 86)	0.96 (n = 253)	268 (n = 40)	3.50 (n = 459)	25.22 (n = 232)	37.85 (n = 40)
Sedentary herds	1.04 (n = 136)	0.91 (n = 76)	315 (n = 22)	3.20 (n = 792)	26.13 (n = 465)	36.82 (n = 43)
Lambing season	ns.	*	ns.	*	***	ns
Winter	1.03 (n = 112)	0.95 (n = 144)	296 (n = 61)	3.17 (n = 552)	24.98 (n = 284)	34.35 (n = 18)
Spring	1.05 (n = 58)	0.88 (n = 157)	285 (n = 18)	3.19 (n = 514)	25.13 (n = 270)	29.28 (n = 61)
Autumn	1.05 (n = 104)	0.96 (n = 180)	294 (n = 25)	3.30 (n = 567)	21.58 (n = 307)	30.81 (n = 25)

Remarks: *, **, and *** indicate significance levels, with: *: $p <= 0.05$, **: $p <= 0.01$, ***: $p <= 0.001$; ns. = not significant.

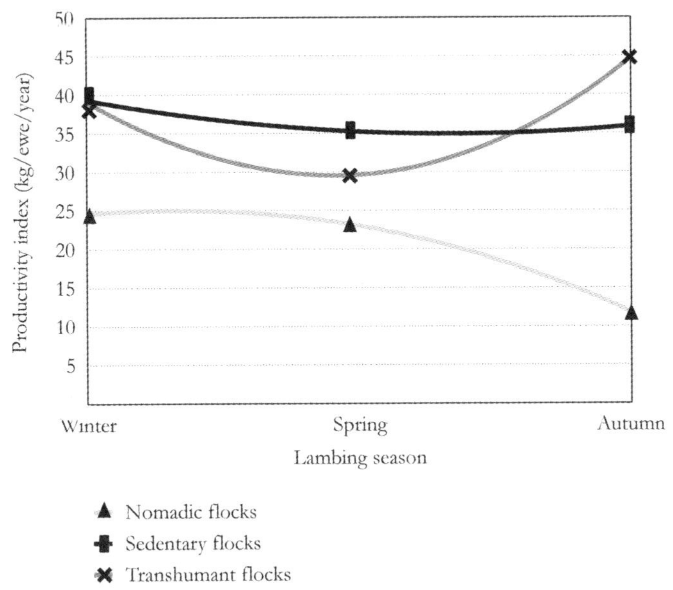

Figure 9-2: Seasonal Ewe Productivity Index for Nomadic, Transhumant, and Sedentary Flocks

the respective livestock producer, as he may, for example, invest in modern water-provisioning infrastructure and thus get more independent on scarce collective water resources.

Effects on Reproduction and Ewe Productivity

Given the three types of sheep production systems, the nomadic, transhumant, and sedentary (cf. Table 9-1), consisting of the named ethnic and income groups, different reproduction characteristics can be distinguished. Table 9-3 summarizes the least-squares-estimates for litter size, survival rate, parturition interval, weight at birth, weight at 120 days and productivity index of the ewes as affected by production system and lambing season. No significant differences in litter size were found between the nomadic, transhumant and sedentary herds. However, the production system affected significantly ($p<0.05$) the survival rate and the lambing interval, and very significantly ($p<0.001$) the weight at birth, the weight till weaning and the productivity index. The highest survival rate till weaning (0.96) was obtained for the lambs of the transhumant herds, while the lowest rate (0.91) was recorded in sedentary herds. The longest parturition interval (315 days) was recorded for sedentary herds, while the ewes from transhumant herds lambed at much shorter intervals (268 days) than their counterparts in nomadic and sedentary flocks. Nomadic herds have the lowest productivity index, namely 19.78 kg/ewe/year. This differs markedly ($p<0.001$) from that of the sedentary and transhumant herds (36.82 and 37.85 kg). The recorded higher survival rate and larger birth and weaning weights can aptly explain the higher productivity index of the transhumant and sedentary herds (cf. Fig. 9-2).

The nomadic herds had the lowest productivity index of 11.77 kg for the lambs born in September, October and December. This productivity index differed significantly ($p<0.001$) compared to that of the sedentary and transhumant herds (35.91 versus 44.75 kg) for the same season. The sedentary herds recorded the lowest fluctuation in the productivity index between seasons (4.14 kg). The nomadic ewes produced twice as much in winter and spring than in autumn and the transhumant herds 14 kg more in autumn than in spring.

The lowest productivity values for the ewes that lambed in September, October and November in the nomadic herds are attributable primarily to the low preweaning average daily gain (ADG) of the lambs born in this season, while the sedentary and transhumant lambs recorded the highest value of ADG in this season. The high nutritional value of stubbles which are available in July and August to the transhumant and sedentary flocks – especially for pregnant and lactating ewes – improve animal body weight in the first four weeks (Guessous 1992). Animals in nomadic herds are confronted with scarce and dry grazing and limited water resources. In addition, nomadic flocks face competition over rangeland from the transhumant animals, which stay on the range areas until winter.

The sedentary flocks maintain a constant productivity level of 35–40 kg throughout the year. During the time when the transhumant ewes use the Mijamouene area together with the nomadic herds, their productivity increases from 30 to 45 kg, while the productivity of nomadic ewes decreases from 23 to 12 kg in the same period. Transhumant shepherds who use the rangelands in addition to other resources gain higher returns from sheep production than the nomadic shepherds. The nomadic shepherds who depend on the rangelands throughout the year are disadvantaged in terms of acquiring and using water resources, crop by-products and forest areas.

The transhumant herds rejoin the Mijamouene area in spring (cf. Table 9-1). Mechanization of animal, water and fodder transport is an important aspect allowing for flexibility in the transhumant system. The transhumant shepherds also have more social and political power in the traditional framework. They therefore can establish better infrastructure, including water cisterns in the rangeland area, and have access to crop residues and feed. They establish themselves in the more favorable parts of the communal rangelands, and thus can use collective resources more inten-

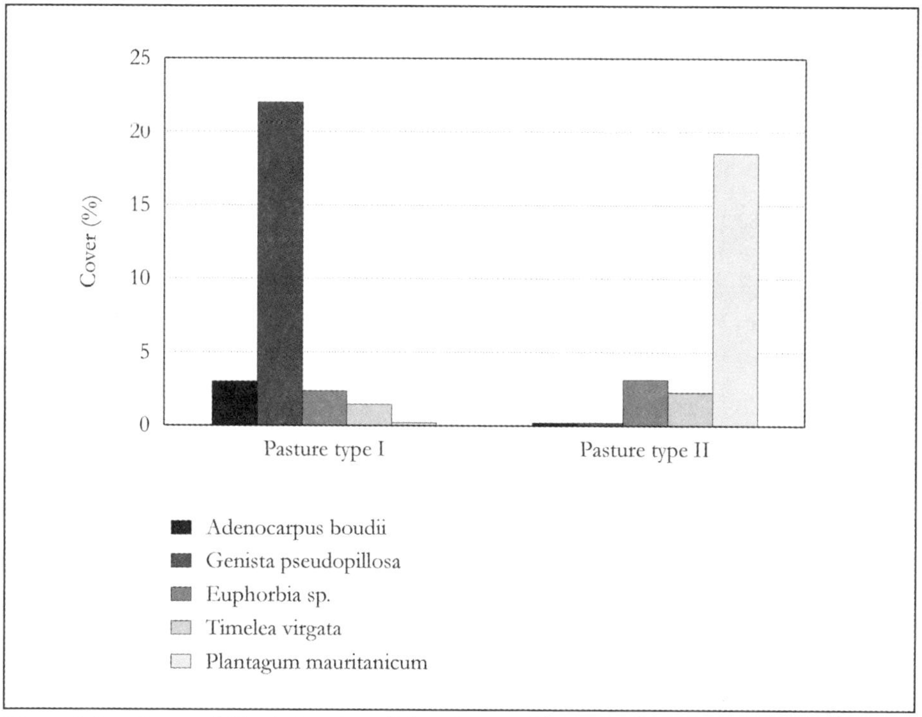

Figure 9-3: Cover of Shrub Species of Two Pasture Types in the Mijamouene Collective Area (in percent)

sively. By strategically complementing various natural resources and inputs, the transhumant system aims to maximize returns (Manriqueet et al. 1996). This has also led to a continuous growth in the size of transhumant herds.

Due to the increasing stocking rates and the limited public water resources at Mijamouene, the nomadic shepherds are pushed increasingly to those parts of the collective range with lower fodder value. There is a considerable potential for conflicts with transhumant shepherds, particularly in periods when the climatic and ecological conditions limit fodder and water availability. The analysis of the fodder calendars and feeding strategies of nomadic shepherds reveals a low level of feed supplementation and infrastructure.

Effects on Range Primary Production

Several parameters point to a trend of rangeland degradation in the Mijamouene area. The first is that in more than 25% of the surveyed areas *Genista pseudopillosa* and *Adenocarpus boudii* have disappeared. These species play an important role as

Table 9-4: LS-Means and Standard Errors for Effects of Interaction Between Pasture Type and Season on Primary Production (kg/DM/ha) ($R^2 = 0.46$)

Pasture type	Season			Ø Pasture Type
	Spring	Summer	Autumn	
Type I	283.12 ± 12.00 (n= 86)	83.68 ± 12.52 (n= 130)	88.93 ± 10.33 (n= 164)	151.91 ± 8.75 (n= 380)
Type II	305.50 ± 16.31 (n= 3)	20.66 ± 19.8 (n= 80)	100.72 ± 19.57 (n= 86)	142.30 ± 12.98 (n= 204)
Ø Season	294.31 ± 10.12 (n= 124)	52.17 ± 11.71 (n= 210)	94.83 ± 11.06 (n= 250)	113.66 ± 98.35 (n= 584)

fodder, because they are able to grow during the winter as well as during the dry season, when the herbaceous and grass species are not available (Djoudi 1998). Because of big canopy surface and deep root growth, these shrubs protect the soil against water and wind erosion and contribute efficiently to improving soil fertility and pasture sustainability (Le Houerou 1993; Papanastasis 1996). The eradication of these shrubs can be attributed to their utilization as combustibles in the nomadic households and to the high stocking rate which reduces their natural regeneration. Another indicator for rangeland degradation is the trend of colonization with *Plantagum mauritanicum*, *Timelea virgata* and *Euphorbia sp.* Because of their toxicity these species are unpalatable and can not contribute to the animal ration except in extremely dry state when adult animals can graze dry leaves in small quantities (not more than one percent of the daily mean intake, Djoudi 1998). The species appear around water points as well as near to the settlements.

Due to scarcity of water points in the range area of Mijamouene, 22 sheepherders in autumn and about 44 in spring use the same collective water point, which is fed by natural water accumulation. Because of the topography of this area, rain water flows from the hills and builds, to the center of the range area, the collective water point. The sheepherders adapt the itinerary of the animals to this situation. Their settlements are located mostly in the peripheries of the range area. From there, all shepherds and their animals converge at the water point in the middle of the day (Map 9-2). This 'forced' herd itinerary increases range degradation, evident in the growing surface covered by less palatable species such as *Plantagum mauritanicum*, *Timelea virgata* and *Euphorbia sp.*

Due to the lack of satellite images of the region in the past, it was not possible to compare the actual stage of vegetation to previous observations. However, even without comparison with historical records, a trend of rangeland degradation from pastures with *Genista pseudopillosa* and *Adenocarpus boudii* (type I, slightly degraded) to less valued pastures dominated by *Plantagum mauritanicum*, *Timelea virgata* and *Euphorbia sp.* (type II, severely degraded) is obvious (cf. Fig. 9-3).

The season and the interaction between season and pasture had a statistically significant effect on phytomass production (cf. Table 9-4). One important result is the relatively high phytomass production of pastures with *Genista pseudopillosa* and *Adenocarpus boudii* (type I), which do not differ significantly from pastures with *Plantagum mauritanicum*, *Timelea virgata* and *Euphorbia sp.* (type II).

The exact analysis of the contribution of different species groups (shrubs, grasses and herbs) reveals that more then 90.7% (129.10 kg/DM [dry matter]/ha) of the total production in these pastures comes from annual and perennial grass species (Poa bulbosa, Festuca ivii and Halcus lanatus). The low phytomass production in the summer (20.66 kg/DM/ha) is due to the interruption of the growth cycle of these grass species. In dry years, which are frequent in the North African climate, the phytomass production from such pastures will drastically decrease and the erosion risk will increase. This tendency was observed in other studies in the Middle Atlas (Bouderbala et al. 1992).

Conclusion

Traditional social structures and property rights constitute the framework of rangeland resource utilization in the agro-sylvo-pastoral system of the Ireklaouen. However, present day production systems are not only a consequence of the traditional framework. According to financial, political and social considerations, members of several clans use the flexibility inherent in the traditional system to diversify their sheep production and maximize their profit. The transhumant herds show the highest productivity indexes as their owners have access to a wide range of natural resources, and to infrastructural and technical means and tools. The nomadic herds strongly depend upon the collective property resources, and thus face a considerable decline in productivity during the months when competition over resources with transhumant herds is high.

There is an urgent need to adapt the traditional collective use of the resources to the new challenges in the pastoral community. The sedentary sheep production system converges to an integrated livestock/crop production farming system. Here, the needs are the improvement of feeding strategies with by-products and the introduction of high value crop or fodder species in fallow areas. For the pastoral and transhumant systems the use and management of water points will be of crucial importance for sustainable range management.

The natural rangelands in the Ireklaouen community are slowly degrading, particularly evident in the growing surface covered by less palatable species, which succeed the original vegetation composed of *Genista pseudopillosa* and *Adenocarpus boudii*. Despite a constant phytomass in both pasture types, the feed quality and the sustainability of the collective rangelands deteriorate. Based on the existing traditional structures, participatory approaches integrating all actors could contribute to reorganize the management of natural resources by the Ireklaouen. A considerable interest of several actors in the region exists and can be used as a starting point.

Notes

(1) Grateful acknowledgment must be made of the important contribution of the herders and farmers in the Middle Atlas. The technical support of the INRA *Programme Viande Rouge*, the *Projet Conseil INRA-Viande Rouge* and the DPA Ifrane, is highly appreciated. Financial support was through the *Deutscher Akademischer Austauschdienst* (DAAD) and the *Gesellschaft für Technische Zusammenarbeit* (GTZ).

(2) This question came up as a consequence of France's earlier experiences with the application of colonial agricultural policies in other regions, particularly in Algeria, which represented a 'laboratory' for several colonial agricultural and property right policies. Administrative and academic discourses related to collective lands are until today largely influenced – even implicitly – by the 1919 *dahir* (Bouderbala et al. 1992).

(3) This migration was induced by similar mechanisms as described above, such as annexation of land for French settlements, the legislation of 10 October 1917 related to forest conservation, and the introduction of pastoral movement restrictions. As a consequence, nomadism and transhumance had become increasingly difficult for them.

(4) For further information on traditional arrangements see Djoudi et al. (2000).

–10–

Territorial Restructuring in Agro-Sylvo-Pastoral Systems of Atlantic Morocco

Mohamed Aderghal

The pastoral system in the region of Oulmes is centered on and articulated with the forest. Embedded in ethno-territorial and tribal structures, it has been transformed by successive socio-economic changes in the region and under the infiltration of the global economy. Its spatial organization permits the mobility of herds and humans, depending on seasonal resource availability. The first real ruptures of the traditional structures began with the French Protectorate, and continued in the context of liberal politics adopted by the national governments after independence. Since then, livestock production adapted to the changing socio-economic conditions. Today's pastoral system as a form of societal organization is thus no longer guided by the traditional principles of regulation. Transformations have had three dimensions: Spatially, livestock production has successively been restricted to the least productive lands. At the same time, new boundaries were imposed by administrative reorganization. Socially, a severe selection has occurred among pastoralists. Economically, pastoral products must now face the risks of competition imposed by the market economy. Against this background, the following questions are discussed: What are the interconnections between the new society and its territory in a context marked by the transition of a mobile pastoral system to an agro-pastoral system more bound to the soil? What place does pastoral activity occupy in the spatial organization of production? People of which (new) social groups are engaged in pastoral activity, and what can be said about their livelihood security?

The Evolution Towards a Delineated Pastoral Territory

The pastoral territory which provides the reference system for the identity of the Ait Amar of Oulmes has been the object of multiple constructions and deconstructions since the second half of the 19th century. Pastoral production has been eroded by a series of ruptures and destabilizations, which were in turn steered by new economic principles and political administration. Generally, the Ait Amar territory of Oulmes developed from a spatially mutative tribal territory to a region with determined administrative boundaries. The ethno-territorial structure in the Oulmes region results from a spatial organization that is based on the means of production as well as on the dimension of spatial appropriation and identification (cf. Map 10-1).

The genesis of this spatial structure and its transformations will be explored in the following. Analysis will be based on two kinds of sources: On the accounts of local history expressing the local peoples' perceptions of their own territory, including the circumstances of its perceptual construction; and on the fractional division established by the Protectorate's authorities (Fig. 10-1), which inscribes the territory with fixed boundaries and various divisions of the administered space.

The Territories of the Local Historical Accounts

Local historical reports demonstrate that the interconnections between the ethnic community and its territory have changed their scale several times. In their accounts of tribal history, all of the interviewed persons mention a territory formerly inhabited by their ancestors and situated in the upper Moulouya, the Middle Atlas, and the eastern depression of the central plateaus. In talking about this 'mythic territory', Ait Amar villagers evoke, for example, ties of kinship with villages in the Middle Atlas. These villages might even carry the same names as those in the Oulmes region.[1] These ties can be considered as the result of bygone tribal developments, when the ancestors of the actual Ait Amar families belonged to the Zaian tribal confederation which inhabited a pastoral territory between the upper Moulouya and the mountains of Oulmes. It was probably towards the end of the 19th century when the Ait Amar migrated definitively to the Oulmes region. This disposition meant their separation from the Zaian confederation and the creation of an autonomous tribal entity.

The creation of the territory that serves the present identification of the Ait Amar goes back to the late 19th and the beginning of the 20th century. This was a period marked by local chiefs' ascension to power on the debris of a declining civil society. By this time, the Ait Amar were a tribal entity formed from an aggregate of clans.[2] They occupied, within the Zaian territory, a frontier position in the *azaghar*, the lowland areas traditionally used as winter pasture grounds by the pastoral populations of the Middle Atlas. In effect, the *caids* Moha ou Hammou of the Zaian and Mohand Aqebli of the Ait Sgougou began to establish their power and to impose their rule over the lands of the *azaghar*, one part of which was cultivated by the Ait Amar.[3] The groups presently forming the Ait Amar were then obliged to retreat towards the highlands of the Massif of Oulmes (cf. Map 10-2).

After occupying the massif, small groups settled in dispersed, but ecologically similar territories. These new territorial units spread on the forest highlands.[4] The seizure of these forest landscapes was realized by four groups – Ait Zitchoun, Ait Alla, Ait Ichou and Ait Affan. It was only possible by playing out competing allegiances of the two rival *caids* of the *azaghar*: the Ait Zitchoun and the Ait Ichou paid allegiance to Moha ou Hammou, and the Ait Alla and Ait Affan to Mohand Aqebli. These political affiliations gave birth to two opposing coalitions, which

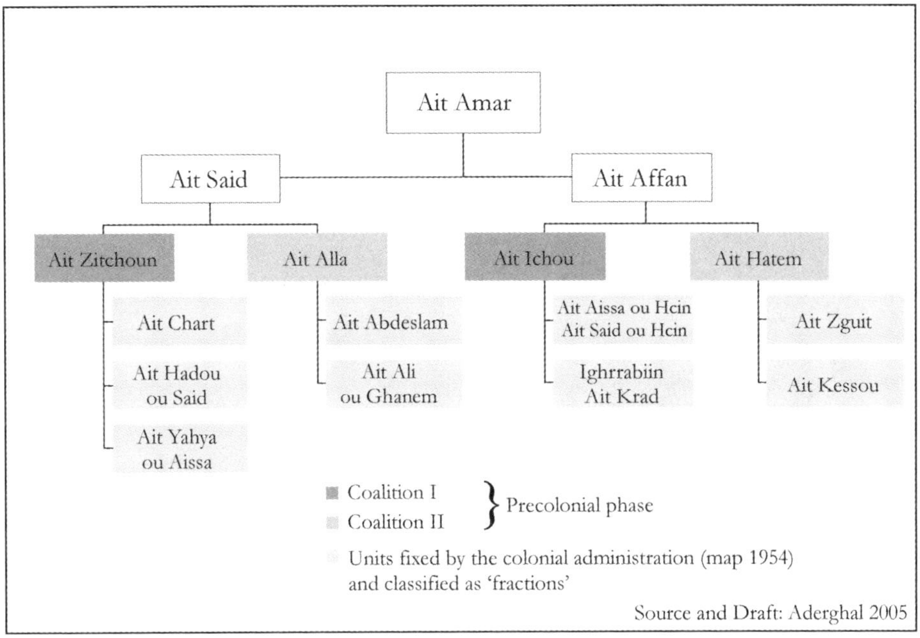

Figure 10-1: Levels of Tribal Structure of the Ait Amar

meant a break of the territorial contiguity. So already in the late 19th century the open territory of the Zaian confederation that stretched across the Middle Atlas and the Massif of Oulmes, began to disintegrate. Each community stemming from the mother confederation established itself on a thoroughly individualized agro-sylvo-pastoral piece of land.

A Map of the Fractions' *Finages*[5]

The boundaries and the configuration of the territory are portrayed on a map dating from 1954. It represents the boundaries of 'fractions' (cf. Maps 10-1, 10-2), a term which in this context connotes spatial units determined by the agricultural tax system, the *tertib*[6]. On the one side, we have the fractions Ait Yahya ou Aissa, Ait Hadou ou Said and Ait Chart. All of them are part of the infratribal group called Ait Zitchoun. Their lands spread out on both sides of the valley defined by the Bouregreg river. They also join the surface terrains of the plateaus of Zguit–Oulmes. These are marginal territories composed of interfluvial vaults. On the other side, the Ait Abdeslam and Ait Ali ou Ghanem, of the infratribal Ait Alla group, occupy lands which stretch out alongside the Oued Tennous, between the plateaus of Oulmes and Telt.

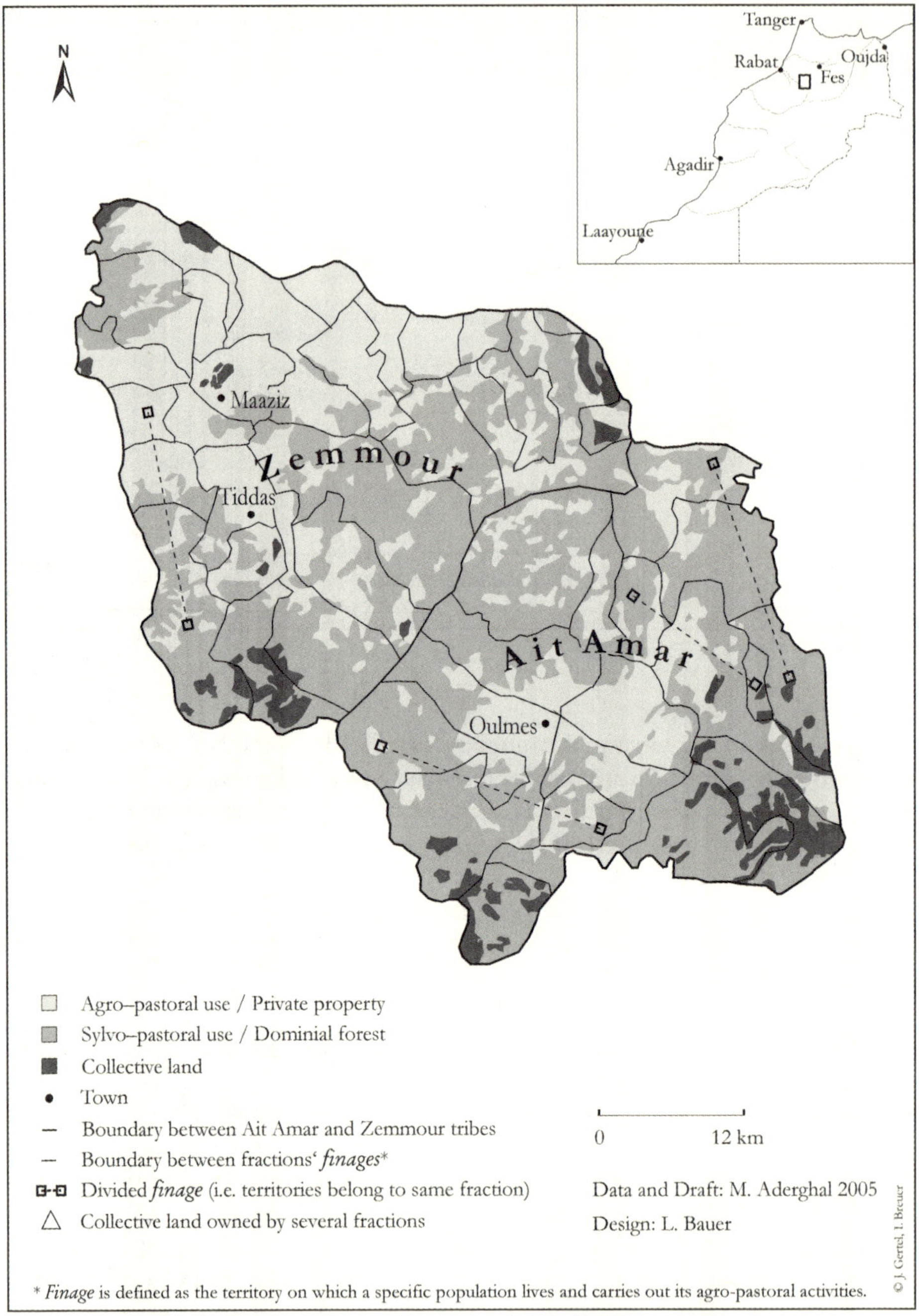

Map 10-1: Massif of Oulmes – Agro-Sylvo-Pastoral Land Use Patterns

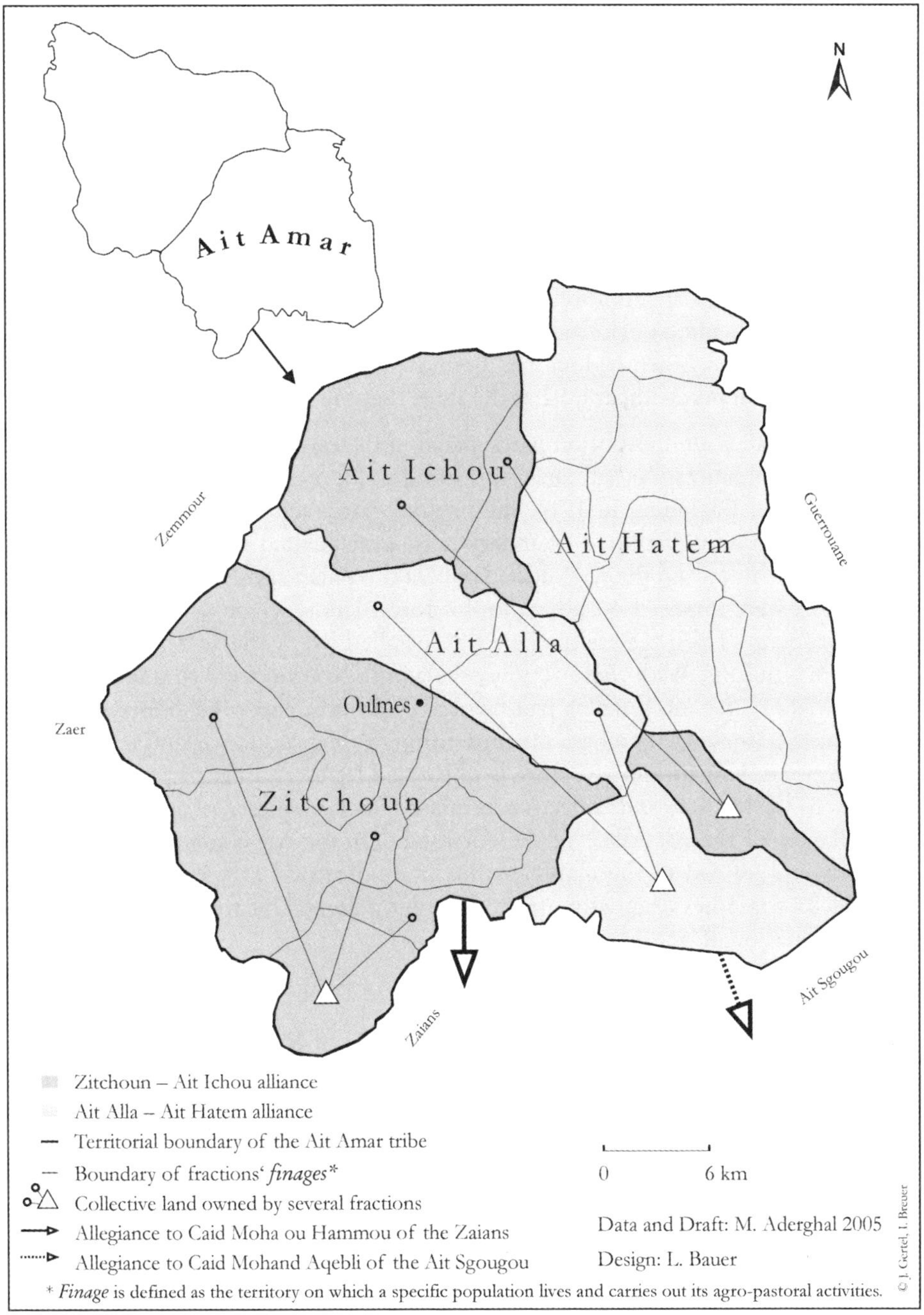

Map 10-2: Ethno-Territorial Organization of the Ait Amar

In the north and east of Oulmes, we have the *finages* of two other groups, namely the Ait Affan and the Ait Ichou, each one subdivided into several fractions. Their territories reach from the Telt Plateau to the basin of Ouljet es Soltane. Among these groups, some fractions possess territorially continuous *finages*, some other fractions' *finages* display territorial discontinuities, especially in the case of Ait Hatem where three or four fractions have bisected *finages*.

This kind of organization of an agro-sylvo-pastoral space in fractional *finages* is often interpreted as being the result of an adaptation of a productive system to the surrounding ecological conditions. It is also suggested that it enables a socially segmented structure and a territorial division of the ecosystem's elements.[7] The analysis of structures in other regions has permitted this interpretation, but I express some reservation herewith, at least at the stage that the map presents.

The protectorate's administration participated in the establishment of this new kind of territorial organization, subjugating local development to a forced administrative regulation. Beginning in 1914, the protectorate's action materialized in three ways. First, an administrative boundary was implemented between the Oulmes-controlled territory and the rest of the lands of the Zaian and Ait Sgougou. Second, restrictions were imposed on utilization of forest lands, which were then declared public property. Third, a state tutelage was installed over collective lands and community institutions. With these actions, the administration tried to undermine the tribal social organization, which had favored coalitions, and which represented the hierarchical framework in which the repartition of the productive land had formerly taken place. This former structure is illustrated by local usage of patrimonial references to the forestry and collective terrains. They are spoken of as forests or collective lands 'of the Ait Alla', 'of the Ait Zitchoun, the Ait Hatem or Ichou'. The fractions represented in the map can thus be understood as socio-spatial entities composed of families originating from different *douars*, knit together by the common exploitation of one ecosystem.

Interterritorial Cleavages and the Emergence of Resource Conflicts

In the heart of these elementary units, recent development has entailed the appearance of new spatial cleavages between the inhabitants of different terrains. The agro-sylvo-pastoral space is organized in connection with the fraction's territory, without border distinction between the *douars* (i.e. the infra-fractional groups). However, the local population settled in small groups of families belonging to different clans of *douars* within the same fraction, or even belonging to foreign clans. Within the same *douar*, the interests of individuals are not identical. Disagreements may arise between individuals, for example between livestock producers settled in forest clearings and agro-pastoralists settled on cropland. These disagreements are often motivated by the unequal access to equipment. Marginalization is not experienced with the same intensity among all groups, and neither do the attitudes towards the forest always depend on the same preoccupations. Some feel directly concerned about the destiny of the forest lands, while others do not see anything there except pasture or a hunting zone.

These communities, united by sharing the same living conditions and the management of the same terrains, have not lost their sense of affiliation. They always feel part of a larger ethno-territorial unit from which foreigners, settled on the same ground, are excluded. This situation is illustrated by the case of pluri-ethnic communities formed around the collective lands of Telt in the small center of Tifoughaline. There, the families identify themselves as members of one of the three ethnic entities Ait Ichou, Ait Alla or Ait Hatem, who in turn distinguish themselves from the foreigners inhabiting that center. The distribution of houses in the center demonstrates these cleavages, since the descendants of each of these three ethnic entities occupy a distinct part of the village. In the nearby *douars* there are cases, where each ethnic group possesses its own mosque. Still, all may speak in unison, for example on matters of forestry management or pastoral resources.

The territorial dynamics and social transformations have led to contradictory interests concerning the restructuring of the communal administration, especially when it comes to the creation of new communal entities which are meant to be closer to the everyday reality of local groups. On the one side, most groups want to make the new entities correspond to their traditional territories. This has an ethnic dimension. On the other side, many groups are made up of families originating from multiple ethnic backgrounds including foreigners, and they justify their claims for a new communal space on the basis of sharing similar living conditions. This can be illustrated by two examples: First, the Ait Ichou have succeeded in detaching themselves from the Ait Affan of Bouqachmir. Since the last communal division of 1992, they have their own commune. Second, the livestock producers originating from the *douar* of Ait Zitchoun are spatially intermingled with workers of the Oulmes water bottling factory who occupy the village of Tarmilet. Both groups suffer from social and spatial marginalization and are thus united in their demand for the creation of a new commune independent from that of Oulmes. Their territory would essentially be composed of forests and rangelands,[8] and its center would be the village of Tarmilet. This demand has thus far not been concretized, for it is not ethnically based, thereby incurring the opposition of influential elites. In effect, some view it as an evil omen that foreigners emancipate themselves in demanding their own commune. Also, if a new commune was created, then the exploitation charges for the mineral waters would cease to contribute to the revenues of the Oulmes commune.

Concluding this analysis, it can be said that the territorial evolution of the region of Oulmes has been dictated by dynamics unfolding in different directions. This paper argues that this is due to forces operating in three different fields, namely economic, political and societal. In the economic sphere, new actors who transgress territorial boundaries have seen their day. The exploitation of resources is realized by industrial enterprises that have become the territory's main actors. So, it is neither the *finage,* nor the communal territory, but what really counts is the territory of enterprise. A symbol for this is the bottling factory, which has created a micro-territory on which the factory workers of Tarmilet depend. Simultaneously, it participates to a great deal in the commune's finances by paying taxes for water exploitation. This allows the enterprise to exercise considerable influence on the decisions of the communal council. Similar mechanisms can be observed at work in other enterprises, for example the tree growing society 'Arbor' a model of plantation economy.

In the political sphere, the state has created powerful instruments for the management of local territories. These conform to a logic divided between two contradictory concerns, namely authoritarian administration on the one hand, and the promotion of development through emancipation of the population on the other. In each case, state action contributes to the territory's organization in submitting the local territory to a technocratic power.

Local society has thus been subject to far reaching transformations. It is now characterized by the presence of social groups which are heterogeneous from the material point of view and belong to different cultural spheres. It has always been a society with an open character allowing the integration of outsiders into the life of the *douars*. In case of social success, outsiders may become influential people and take part in the decision-making process concerning the collective. They may even be elected to the communal council. In contrast to this new influx, the local populations also find themselves increasingly concerned about a process of emigration, to which the first army recruitments and the colonial administration gave birth. The return of these emigrants, after retirement or on the occasion of the annual holidays, is commonly interpreted as a catalyst of socio-cultural change that propagates itself even to the most remote villages. Formerly, these influences were limited to the emigrant's family, and they meant material aid for survival or the amelioration of farming conditions. Presently it can be noticed that expatriates increasingly try to take part in the local administration, either in passing the classic channel of communal elections, or by means of engaging in the recently emerged associative movement. Thus, the territorial organization is no longer directed solely from the center of Oulmes. Rather, it is subjected to multiple management on the part of a multitude of actors with often contradictory strategies.

Specialization of Productive Spaces

The evolution of the pastoral system has been characterized by the extension of farmland and the retreat of livestock production in the whole of the territory. Productive spaces have undergone processes of specialization. This is due to selection processes which depend on the potential of resources available in each territory. As a result, pastoralism is increasingly practiced on the least productive soils.

The pastoral system is constituted by three spatial elements: The forest, the non-forestry pastures, and the farmlands. The transformations mentioned above have contributed to the perturbation of the pastoralists' livelihood systems. Within a few decades the traditional agro-sylvo-pastoralism, where resource use was based on the principle of spatial and temporal complementarity, transformed to a system in which mobile livestock production on unproductive soils is combined with agricultural use of the productive lands.

The original forestry ecosystem has in many cases been degraded to scrub. However, the forest's role in the survival of the pastoral system is essential. In principle it supplies more than half of the stocks' alimentary needs and constitutes an important facet in the seasonal mobility of herds, which on average sojourn in the forests 200 days annually. It is certain that the forest's role in the pastoral system is no longer what it was before 1917, when the colonial legislation restricting access was implemented. The tendency towards a decline of the forest's alimentary contribution to livestock diets can be connected to three facts. First, territories used for forestry have been delineated and pastures have been subjugated to an administrative regulation. Second, the forest area has been reduced following the extension of arable farmland caused by shepherds who recently converted to livestock producing agriculturalists. And third, the economic and cultural function of the forest has been transformed, and it is now a space managed for forestry and open to the recreational activities of city dwellers.

The rangelands only subsist where agriculture is limited by edaphic conditions or constraints such as slope incline. They are often reduced to single patches of callous soils and are exposed to erosion. Those that are part of the collective lands are subdivided either in blocks of some hundreds of hectares, or splintered into a multitude of small parcels surrounded by the forests. Now the characteristic image of vast willows within the pasturelands of the highland of Oulmes is a mere vestige of its former state.

Some of the cultivated lands contribute largely to the local pastoral economy, especially those situated on slopes exceeding 15%, slopes to which farmland has been extended as a consequence of the local paucity of arable land. These lands are vulnerable to erosion, their productivity is low, and they are often designated for use of fallow and gleaning. They fulfill a double function on the pastoral plan: to supply a part of the alimentary needs of the herds; and to serve as a basis for an easy access to the forest.

The most productive soils, however, are reserved for agricultural utilization at high added value. Situated on the plateaus, they have experienced an important land concentration. Private investments have allowed the formation of enclaves of modern agriculture, where systems based on perennial market-oriented cultures are utilized, in particular on the plateau of Oulmes, where more than 500 ha are occupied by arboriculture and more than 300 ha by lavender. On the Telt plateau some hundreds of hectares of an ancient collective pasture are cultivated with cereals by an urban farmer on the basis of a long-term lease contract. Similar examples can be found on volcanic terraces along the valleys or on very steep slopes.

In summary, the economic transformations that occurred since the first years of the Protectorate have reinforced marginalization and contributed to the perturbation of ancient pastoral systems. The motor of these transformations has been the private capital of urban origin and the new entrepreneurs not imbued by pastoral culture. The financial assets invested in entrepreneurial livestock production, via associations, constitute a more abundant reward than an investment oriented towards the modernization of production structures. Furthermore, the partial extraversion of the pastoral space was implemented by non-agricultural activities, such as the industrial and commercial exploitation of the mineral waters of Oulmes, the forestry resources, and some mines. The spaces dominated by urban capital remain, however, isolated spots in the landscape, and their organization is based on the tapping of natural resources and workforce. Nevertheless, they constitute the genuine vectors that integrate the region into the global economy.

Growing Economic Inequalities – Indicators of a Social Crisis?

Social marginalization affects a substantial part of the local population. This, as I argue, has to be seen in connection with the spatial transformations described above. Poverty is a consequence of spatial injustice. The most vulnerable groups are the smallholding farmers, who are incapable of confronting the constraints posed by scarce pecuniary means and those of a physically demanding environment, where droughts are not uncommon.

Though the phenomenon of poverty is not new in the region, the understanding of it is still at its beginning. The lack of precise and pertinent data renders its evaluation delicate. In the beginning of the 1980s a classification of enterprises in the region, undertaken by the Oulmes-Rommani Integrated Development Project, made the distinction between three types of farms, based on their monetary revenues:

First, the small marginal and subsistence enterprises constitute a category composed of landless farmers (workers, shepherds, herd owners or associates) and owners of less than 4 ha of land. They represent 37% of the population. The per-head revenue in this category was about 100 to 120 USD, which is well below the threshold of absolute poverty of 224 USD (895 Dh), as defined by the World Bank for the 1970s.

Table 10-1: Vulnerability Indicators of the Rural Populations in Oulmes

	Poverty rate (%)	Severity of poverty (%)	Vulnerability rate (%)	Average expenses (Dh/head)
Rural communities				
Oulmes	15.0	4.4	40.5	8,711.8
Ait Ichou	38.9	14.2	59.9	4,863.2
Bouqachmir	43.2	16.9	62.1	4,757.0
Comparison				
Province	22.1	7.3	49.0	5,472.0
National	22.9	5.5	51.1	7,008.7

Source: HCP (2005).

Remarks: The term 'poverty' refers to monetary poverty. A household is defined as poor if expenses are below the poverty line. For rural Morocco, poverty lines are defined as follows (cf. the 1998/99 *Enquête nationale sur les niveaux de vie* [National Survey on Living Standards]): 3,037 Dh per person and year for relative poverty, 2,642 Dh/p/y for absolute poverty. The column 'poverty rate' indicates the percentage of individuals in the respective administrative entity whose expenses are below the poverty line. In the column 'severity of poverty', the percentage of the poorest among the poor is indicated. A household is defined as being economically vulnerable, if its expenses are inferior to 5,883 Dh per person and per year (urban) and inferior to 4,555 Dh/p/y (rural).

Second, the semi-commercial farms have surfaces ranging from 3.5 to 31 ha. They occupied 68% of the arable land and obtained 70% of the agricultural production. 60% of the rural population lived from revenues of these farms, and their annual income reached 350 USD per head (1,000 Dh).

Third, the large commercial farms are real market-integrated enterprises. These are the farms that appear with an enormous concentration of lands. The largest farms reached 400 ha at Oulmes, which is much compared to other areas of the wider region (Rommani: 167 ha; Ezzhiliga: 93 ha).

In 2004, the map of poverty established by the *Haut Commissariat du Plan* regarding the rural populations in the communes Oulmes, Bouqachmir and Ait Ichou gave the indicators presented in Table 10-1.

By comparing these two sets of data we can gain the following insights into the structure of poverty in the region: In the 1980s the average per-head revenue did not exceed 250 USD (1,000 Dh), for more than 60% of the agro-pastoral population. More than 37% lived below the absolute poverty level of that time period (224 USD per head). For 2004, we can observe that poverty is less salient in the community of Oulmes compared to Bouqachmir and Ait Ichou. In these latter communes, not only poverty rates are high, but also the vulnerability rate approximates 60%. This is a much higher proportion than that registered at the provincial or regional level. The data displayed here thus show extreme disparities and profound socio-spatial disequilibria within the Oulmes territory.[9] They also reveal that in several communes misery has been aggravated during the last decades.

Within these general patterns of poverty, pastoralists are among the most vulnerable and needy. The results of a survey[10] that we have conducted in several *douars* of the communes of Oulmes, Ait Ichou and Bouqachmir allow us to present the following conclusions: The household heads who exercise an activity as agro-pastoralists constitute a majority that in certain *douars* may comprise even all households. Workers and shepherds are also predominant in certain *douars*. That is the case at Ait Ben Azouz close to the center of Oulmes and at Tifoughaline, where they represent, respectively, 50% and 40% of all households. The economic activities in these *douars* still depend on the agro-sylvo-pastoral system. Animal husbandry is an activity that continues to be fundamental for the large and medium-scale landowners. For the more needy it is an activity, which is only possible if they have access to the forest. But this is becoming more and more restricted due to purchases by the more affluent. This tendency intensified during the recurrent droughts of the last decades.

In the wake of the transformations discussed above, the social groups in the *douars* are not all exposed to the same degree of vulnerability. At the same time, we can identify several groups whose social conditions are more or less dependent on revenues from pastoral activities. The qualitative information that we have collected in course of our different surveys permits us to advance to a first draft of the social profiles of these groups.

The large landholding livestock producers: They generally head farms of more than fifty hectares and flocks of hundreds of animals, composed of sheep and cattle. They enjoy a social status of local nobility. Normally, they do not directly perform pastoral work, rather they delegate the daily shepherding tasks to their familial helpers, tenants or salaried workers. Their life is divided between the *douar*, the local center and the city, and the sources of their revenue are not exclusively agricultural. This makes them relatively invulnerable to the risks generated by the fluctuation of agro-pastoral revenues.

The medium landowning livestock producers: The owners of middle-sized farms are less well-equipped with land. They own between five and twenty hectares. They succeed in building up important herds thanks to the institution of shareherding associations and the dispersion of flocks on different pasture grounds. Nevertheless they remain vulnerable to socio-economic risks and to natural constraints that burden livestock production, especially when there is no secondary source of income or no emigrated family member whose money transfers allow for a balanced budget.

Both the large- and medium-scale landowning livestock producers continue to practice livestock production in a traditional manner, based on natural collective pastures and forestry terrains with little integration into the cultivation system. Stall-feeding of cattle and even sheep is omnipresent, but it concerns only a very limited percentage of the herds. It can be considered as a practice depending on temporary demand fluctuations, but it does not alter the pastoral structure as a whole. Neither does the orientation towards dairy farming, initiated by some notables in Oulmes. These structures find their explanation in the disposability and the gratituity of the resources on which the production system is founded. Extensive livestock production, accompanied by an increase in the number of salaried shepherds, allows the animal owners to draw the maximum profit of pastoral resources furnished by the forests.

The small landholding livestock producers: With farms smaller than five hectares, and even not exceeding one hectare for the large majority, this group lives at the limit of subsistence and is consequently very vulnerable. If a farm is located close to a forest, the farmer may be able to take stock in association at a fifty-percent-share – instead of a twenty-five-percent-share that he would be paid if he was landless. But the productivity of forestry lands is in decline and the livestock producers are obliged to buy fodder on the markets or to relocate towards better pastures. The repercussions of fodder expenses are felt especially during dry years, where scarcity of water holes adds to the alimentary penury. Weak herd mobility is a signal of the demise of the pastoral activity, which is no longer profitable for owners of small parcels. This is especially so if financial means to support the herd's alimentary needs and health requirements are insufficient. The abandoning of livestock production announces the beginning of a process of uprooting and of search for other means of surviving. The pastoralists concerned may resort to the illegal exploitation of forestry resources, chopping wood or producing charcoal, which they sell in the closest centers, like the wood cutters of the *douar* Benazouz at Oulmes or the Thuja-fellers of the *douar* Iktfane at Bouqachmir. These illegal activities complement revenues earned by agro-pastoral activities, and thus enable pastoralists to keep a residence, be it only temporary, at the *douar*.

The landless livestock producers: The landless constitute a social group whose livelihoods depend on activities as family auxiliaries, salaried workers or stock-keeping associates. 'Family auxiliaries' are family members – married or not – who take care of the family's flocks under the authority of their father. They are shepherds who are provided for by the family and who may gain small additional revenues by odd jobs or by engaging in an associateship for livestock production. If they are affiliated to families of medium or large landowners, this gives them the advantage of relative stability in their relationship of dependence, while they await the disappearance of their father in order to access the property and become patriarch and landowner. Often, their path of life is suddenly altered by ruptures that may occur within the family. Landless shepherds – salaried or associated – are occupied with guarding and conducting the herds that are entrusted to them by the medium or large-scale landowners. Salaried shepherds are generally young and single. In contrast, associate shepherds tend to be married family fathers possessing their own tent or settling in a hut on the landowner's property. This type of shepherd is called *amakssa n'rebaa* in the *tamazight* language, which literally means "shepherd with twenty-five-percent-share"; the landlord may accept a certain intimacy with him, sharing meals or receiving him under his roof. In other cases, the shepherd may be held at a distance.

If the *rebaa* is a stranger, he rarely accedes to the status of a landowner. The only possibility that he has after a long life of privations and solitude, is to press for a fifty-percent-share or to possess his own livestock. That means that in the pastoral society the *rebaa* occupies a paradoxical social position. Due to his deep knowledge of livestock production, he is integrated into the professional world, but because of his modest revenues, he suffers an economic and social marginalization. So, livestock production is barely profitable for those who directly perform the shepherding activity. Their material situation is almost always miserable. This is not only the result of a decline of livestock production and of the diversion of a part of its added value to the profit of mediators and butchers, but also of a traditional perception that turns the noble landowner into a rent seeker and the shepherd or the associate *rebaa* into a simple servant.

Conclusion

The recent restructuring of the territory in the region of Oulmes has not been sufficiently effective to permit the overhaul of the structures that have been inherited from local history, namely the tribal and colonial periods. And this in spite of many decades of an opening towards the influx of people, goods and ideas. The ethnic discontinuities and the disparities between unequally developed spaces continues to strongly influence the territorial organization in the region. The pastoral system as a form of internal organization of a local society is no longer guided by the same principles of regulation as formerly. Adapting to the new socio-economic context, livestock production, even if it remains a productive activity, no longer constitutes a means for social integration. The restrictions on the utilization of open pastoral space pressured towards a reduction of herd mobility. This reduction is accompanied by growing economic inequalities, the specialization of productive spaces, and resource conflicts.

Notes

(1) One example is the village of Ait Ali ou Ghanem in the Aghbala region and a village with the same name situated in the Ait Alla fraction of Oulmes.

(2) The Ait Amar tribal entity had already existed for a long time. They are, for example, mentioned by the historiographer Addu'ayyif in his account of an event that took place in 1810. In that year, a convoy of prisoners was conducted to Sale by the governor of Boujaad; the prisoners belonged to the five clans Ait Ichou, Ait Alla, Ait Hatem, Ait Boumet and Ait Ihichou, all of them part of the Beni Amer (Ait Amar) tribal fraction of the Zaian (El Amari 1986, 351).

(3) An account of the same events can be found in Abes' note about the 'Ait Amar d'Oulmès', Berber Archives 1916.

(4) Between the high valley of Oued Ksikou (1,066m) in the south, the basin of Ouljet es Soltane in the north, on the opposite side between Jebel Adar n'Hyanes on the Telt Plateau (about 1,300m) in the east, and Jebel Harcha in the northeast (about 1,000m).

(5) The term *finage* is defined as the territory on which a defined population lives and carries out its agro-pastoral activities.

(6) Tax on agricultural outputs and livestock.

(7) This is an interpretation very dear to geographers which, as I suggest, projects the notion of altitudinal complementariness from the context of the Alps on the Moroccan mountains.

(8) On the Western borders of the plateau of Oulmes and on the plateau of Zguit.

(9) In the case of the Oulmes community, the poverty indicators are, however, misleading. This is mainly due to the fact that the agglomeration of Oulmes is classified as urban and thus not considered in the calculation. Concerning communal revenues, Oulmes is one of the richest rural communities of Morocco, mainly because of income from the forests and from the Oulmes and Sidi Ali mineral water factories.

(10) The survey was undertaken in 2003 within the framework of the *Programme de développement rural intégré des zones forestières et périforestières* (Program of Integrated Rural Development in Forestry and Periforestry Zones) for the *Haut Commissariat aux Eaux et Forêts* (High Commission for Water and Forests), in collaboration with M. Ait Hamza.

–11–

Livelihood Security and Mobility in the High Atlas Mountains

Ingo Breuer

Hammou, 68 years old, lives with his family in a mudbrick house in the village of Taboulmant in Oussikis. He has two sons. The elder, Lahcen, is married, and he is working on a daily basis in the construction sector. Each year, he spends between four and eight months in the city of Casablanca. Hammou's second son, Moha, is married as well. He identifies himself as nomad (arhhâl). He takes care of the family's herd, some 120 sheep and goats, and he lives in the tent throughout the year. In summer, he stays on the pasture of Idikel, about 15 km from the village. In winter, he might transport the animals to pastures situated between 100 and 700 km from Oussikis, either to Jebel Saghro, or to the Souss region in the southwest. Both of Hammou's sons leave their wives and children in Oussikis during the whole year. One exception is Moha's nine year old son who helps his father to take care for the animals.

Hammou is the head of this household. He is responsible for the family budget, and he shops at the weekly market for all family members. Hammou also makes all major economic decisions, for example, which animals to sell at which moment, or which amounts of money to invest in the different activities. In summer, when Moha locates his tent in Idikel, Hammou will commute between the tent and the house. Otherwise he cultivates, together with the women, the few family fields on which they grow wheat for self-consumption.

During some years, the family's budget is based entirely on the sale of animals. In years of drought, on the contrary, the family lives almost exclusively on what Lahcen earns in Casablanca. Sometimes, they even use some of this money to buy barley as supplementary animal fodder to prevent the animals from wasting. But this strategy can only be successful if Lahcen manages to find a well-paid job. In 2001, for example, he went to Casablanca, searched for weeks and didn't find anything, so that he had to return without a single dirham earned. In years when all of these strategies fail, and neither wage work nor the sale of animals will make a living, the family has another source to tap. They can rely on financial support from an uncle who emigrated to France in the 1970s and whose donations are a recourse in years of drought.

The present paper discusses pastoral economies and strategies in the High Atlas mountains. It is based on a study conducted in the Oussikis valley, located at the southern slope of the central High Atlas, between 2001 and 2003.[1] As many other valleys in the Atlas, Oussikis is characterized by deep social inequalities: Some pastoralists can rely on a stable livelihood, while others live in social insecurity and poverty. It will be argued that these inequalities cannot be explained by simply comparing the animal production systems of the respective households. There are two reasons for this. First, livestock production in the central High Atlas almost always takes place in the context of multi-resource and multi-activity household economies. Second, decisions concerning the herd economy are influenced by factors from outside the pastoral sphere, for example, the ability to enter urban labor markets. An evaluation of pastoral perspectives thus must consider the fact that pastoralists' activities are always embedded in broader livelihood strategies which transcend the pastoral sphere. The present paper thus aims to investigate how pastoral livelihood strategies are shaped by unequal asset and resource endowments on the household level.

Oussikis – A Socio-Economic Setting

Oussikis is a large valley, about 2,100 m above sea level, situated in the southern part of the Central High Atlas. It is part of the community (commune rurale) of Msemrir, in the northeast of the province of Ouarzazate. The valley floor, alongside a perennial river, is used for irrigated agriculture. The valley is surrounded by wide pasture areas which make up more than 99% of the community's territory. Oussikis is inhabited by a *Tashelhayt*-speaking tribal fraction identifying themselves as Ait Oussikis, literally: the people of Oussikis. They are comprised of several groups of the former tribal federation of the Ait Atta. The case study covers five villages in the upper part of the valley, inhabited by three different Ait Atta groups, and with an overall population of about 4,200.

In the pre-colonial era, the local economy was based, to a large extent, on mobile livestock production of sheep and goats. The Ait Oussikis followed a transhumant mobility pattern between the summer pastures of the High Atlas and the winter pastures of the Jebel Saghro and the Pre-Saharan plains. Nevertheless, they were never entirely nomadic, but combined livestock production with irrigated agriculture, and they tended to stay seasonally around their villages where a part of the family used to live in houses and looked after the fields. Until the 1940s, this economic basis had not fundamentally changed, though major disturbances of the transhumant system had occurred in the context of the French intrusion which had reached Oussikis in the early 1930s.

Since the late 1940s, Oussikis has been subject to slow, but far-reaching economic and societal transformation processes (Breuer 2007). Three aspects are crucial for the understanding of today's local economy: First, labor migration, both temporary and definitive, to the thriving Moroccan cities as well as to Europe, has become a central source of income. Subsequently, the Ait Oussikis have built up complex social networks linking the valley with urban Morocco and even Europe. Simultaneously, many households in Oussikis have adopted 'multi-local' livelihood strategies: They may comprise mobile pastoralists, 'sedentary' persons living in the village, and persons who work periodically in the cities. This context shapes pastoralism as it is conducted by the Ait Oussikis today. An analysis of the local pastoral system can in no way be separated from this setting.

Table 11-1: Economic Activities of the Male Adults of Oussikis

Main activity	Percentage of persons
Wage labor in informal sector	44.4%
Mobile livestock production	14.7%
Self-employed in trade, services, artisanship	10.5%
Self-employed as construction entrepreneurs	3.3%
Agriculture	8.4%
State-employed	5.4%
Emigration to Europe	5.1%
(Other)	1.3%
(No activity)	6.9%
Total	100.0%

Source: Own survey (n= 905). All men of Oussikis, between age 16 and 64, are included.

Second, the study revealed that the local economic system has moved away from reliance on livestock production, and that it is characterized by very diverse income sources. Its structure is displayed in Table 11-1, which represents the economic activities of the male adults: Almost half of them (44.4%) are engaged in informal wage labor. 14.7% of them are mobile pastoralists. Apart from these two groups, there is a group of self-employed people exercising diverse professions (13.8%). Some of the Ait Oussikis are state employees, and there is also a group of emigrants who work in France. Those who declare farming as their main activity are comparatively few (only 8.4%).

Third, spatial mobility is thus an important feature in the everyday life of the Oussikis men (cf. Map 11-1). Only about 40% of them live permanently in Oussikis. All of the others are mobile, be it as nomads on the pastures (14.7%), as emigrants in Europe (5.1%), or as temporary labor migrants in cities as Rabat, Casablanca, Marrakech, Ouarzazate, Agadir, and the Sahara (39.0%). The local economic system is thus tightly knitted to urban labor markets through work force mobility.

Those who are active in informal wage labor, work almost exclusively as temporary work migrants in the cities. About 90% of them are employed in the construction sector, as low- or medium-skilled workers. Most of these are 'temporary migrants' in the literal sense of the word. They stay in the cities for between four and ten months per year, leaving their families in the village. The majority of them returns to Oussikis for at least two or three months per year. Many of them stay away from Oussikis whenever they find a job, but they may spend long periods at home whenever they do not find one. This 'temporality' is partly the effect of the structure of urban labor markets in the informal sector, especially concerning low-skilled employments in the construction sector. Wages are often too meager to support an entire family in the city. Thus, the migrants often live on the construction sites, in tents or garages. But the major factor is perhaps that, in order to find work, the migrants have to be mobile, always ready to go to wherever they might find work for a few weeks. In 2002, 31% of the labor migrants of Oussikis had worked in at least two different cities. Under such circumstances, it becomes obvious why comparatively few of the Ait Oussikis families have definitively emigrated (82 families in the last 30 years). However, even these families are part of the extended network the temporary migrants arriving in the city can rely on.

The Ait-Oussikis-entrepreneurs hold a special position in Oussikis, as they play a crucial role in the local economy. There are about a dozen entrepreneurial families in the valley. All of them display very complex family histories in which, over several generations, spatial mobility, emigration, material investments in different cities, and investment into education, were closely intertwined (Breuer 2007). In the recent decades, these families have constituted large enterprises mainly in the construction sector, and they are now active in nearly all Moroccan regions. Though operating nationwide, they stay locally based, recruiting a large number of young men from the valley to work in the construction sector. And it is certainly in this form of 'social capital' that the work force mobility of the Ait Oussikis is rooted.

Oussikis has an irrigated agriculture of mainly wheat, barley, potatoes, luzerne, and apple trees. Almost all households (96%) own land, and in virtually all of them we can find subsistence-oriented agricultural activities, mostly wheat and barley production, and sometimes small-scale cash-cropping of apples. At first glance, agriculture seems to be thriving. It has experienced a great deal of modernization since the early 1980s, with the construction of a large barrage and the massive introduction of apple trees. However, population has tripled within the last four decades, and as heritage is normally divided among the sons, land has become scarce in many households: More than 50% of the households now own no more than one or two little terraces. Additionally, since the 1980s, a small class of ex-emigrants and entrepreneurs have invested in land in order to create large apple plantations. Land tenure is thus becoming more and more polarized, and today many households are no longer able to produce enough cereals for their subsistence.

Irahhaln n-Ait Oussikis – The Oussikis Pastoralists

Oussikis pastoralists identify themselves as *irahhaln* (sg. arhhâl), a term conventionally translated as 'nomads'. The *irahhaln* live in tents for most of the year, and sometimes also in caves, but they all have houses in their respective village. They keep sheep and goats on collective pastures. In the majority of the cases, livestock production is not the only economic activity in their households. Almost every *irahhaln*-household – like each household in Oussikis – is characterized by a specific combination of economic activities. One person, for example, may be *arhhâl*, the others may pursue non-pastoral activities such as labor migration, or they may stay in the village all year. The case study of Hammou cited above illustrates a typ-

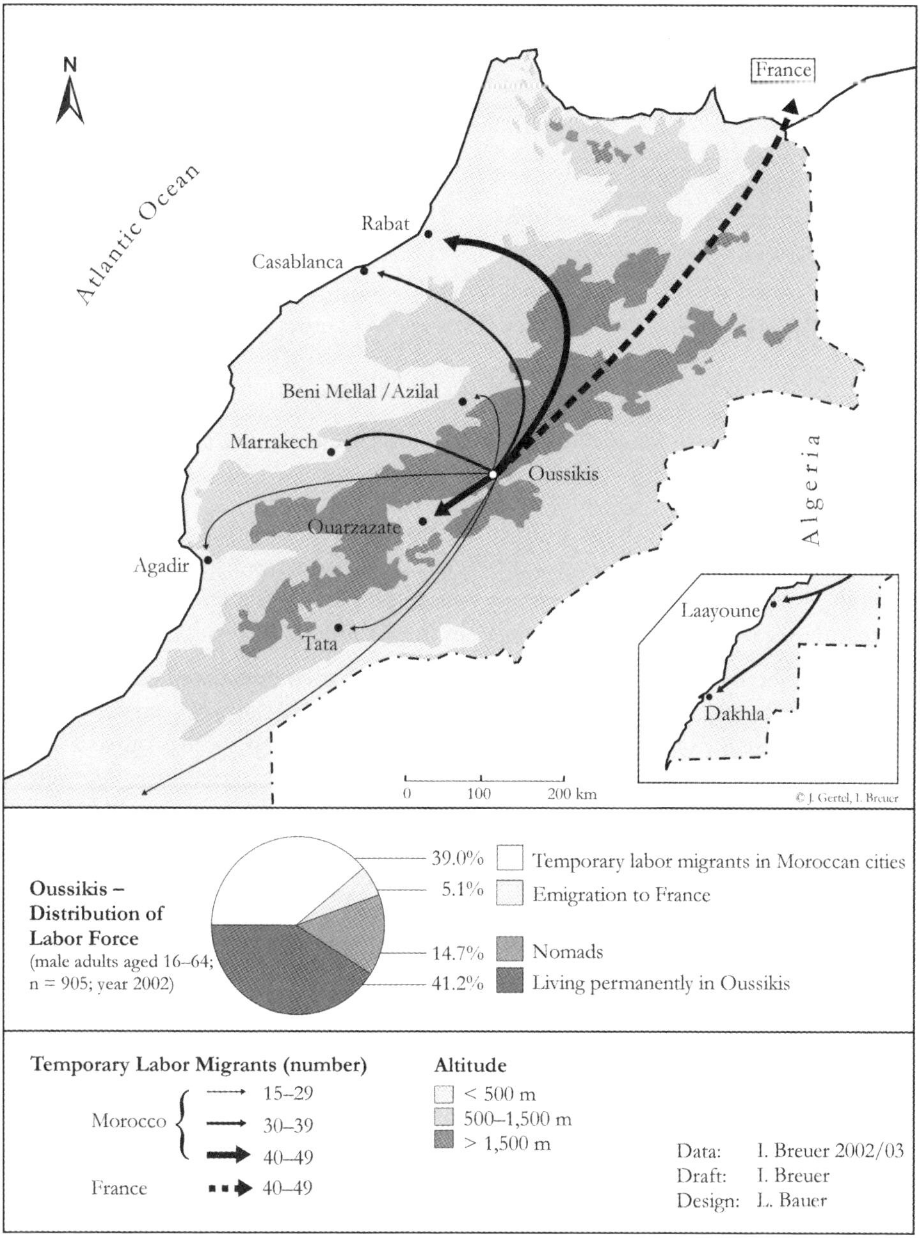

Map 11-1: Labor Mobility in Oussikis

Table 11-2: Types of Pastoral Households in Oussikis

Type	Activities of male adults (agricultural activities not represented)			Percentage of households [n= 82]
A	Pastoral activity			30.5%
B	Pastoral activity	+ Wage labor		39.0%
	Pastoral activity	+ Wage labor	+ Self-employment	12.2%
C	Pastoral activity	+ State-employment	+ >2 other activities	11.0%
	Pastoral activity	+ Emigration (France)	+ >2 other activities	7.3%
Total				100.0%

Source: Own survey.
Remarks: The 82 pastoral households represent 22% of all households in Oussikis.

ical structure of *irahhaln*-households: In many of them, livestock represents only a fraction of the portfolio of several different income sources. One has thus to consider the varied household contexts in which the pastoralists are involved, as well as the role the pastoral activity may play in these different 'reproductional settings'.

Pastoralists' activities are integrated in differently shaped household economies (cf. Table 11-2). In the study area, 22% of the households (82 of 373) pursue pastoral activities. These *irahhaln*-households can be subdivided into three groups: Type A (25 of 82) refers to households that, except for small-scale agriculture, specialized on their nomadic activity. The sale of animals is their only income source, apart from very irregular situations in which they may resort to short-term supplementary strategies. Type B (42 of 82) are households in which pastoral activities are regularly combined with informal wage work, and sometimes additional self-employment. These households comprise several male adults some of whom are nomads, while others work in the construction sector. Type C (15 of 82) are those households, which in addition to their pastoral activity have access to formal employment or to international migration. At least one of their members is a state employee or emigrated to Europe.

In about 70% of the *irahhaln*-households, herd keeping is merely one of several activities. The case of Hammou illustrates only a few of the complex articulations between the 'pastoral sphere' and other economic spheres. It also reveals that pastoral livelihoods are very much determined by livelihood factors, such as the availability of family work force, the ability to enter urban labor markets, the degree of income diversification, or the capability to access reliable social networks. Pastoral activities are also embedded in multi-local strategies. Many of the households are literally split up between the village, the tent, and, in many cases, Moroccan cities or even European countries. With respect to this, the case study revealed another feature typical for Oussikis and the High Atlas: the Ait Oussikis have maintained a unique system of household coherence, where households may be scattered to several places, but resources are managed collectively. In such a socio-economic setting, an analysis of pastoral livelihoods requires an analysis of the resource endowments at the household level.

Resource Analysis: Evaluating Pastoral Livelihoods

In the following, a livelihood analysis of the 82 pastoral households of Oussikis will be presented, in undertaking a cross-sectional analysis of their resource endowments. The following questions will be discussed: Which households dispose of what kinds of resources? Which ones are vulnerable and which ones less so? What role does the herd play in the economic system of the different households?

The base of the analysis is a conceptual framework distinguishing between four basic categories of resources. It is based on the works of Gertel (1997; this vol.), but adapted to the local conditions of southern Morocco, by operationalizing it into 12 variables, three of which are assigned to each of the four categories (Table 11-3):

Incorporated resources are assets which are immediately bound to the body of a person. Three types are distinguished: (1) Work force, that is the age dependency ratio (ADR) of the household, defined as the number of working-age men in relation to the number of children and old people; (2) Education, i.e. formal school degrees that may open up access to state jobs; and (3) Health, or the absence of chronic illness (illnesses being a substantial burden for a household's economy).

The term socially institutionalized resources (SIR) is used for assets emerging from social relations or group affiliations through which claims can be made. Here, the following stratification is applied: 'International SIR' mean contacts to kin located abroad which provide material transfers, be it permanently or in times of need only, like Hammou's uncle. 'State SIR' include claims to state assistance, such as social or health insurance. These claims are important as they protect the household from risks, such as debt in case of accidents or illness. 'Local SIR' means that a household counts local officials among their close kin. In the local context, additional resources may me mobilized by means of these relations.

The term allocative resources refers to the physical assets a household possesses. Three types are included here, namely: (1) Productive goods, that is agricultural land, trees, and animals; (2) Material goods, which include a selection of household items indicating wealth (such as bicycles, kitchen devices, television sets, and access to electricity); and (3) the household's capacity to produce food (here cereals) for its own consumption.

Table 11-3: A Scheme for Quantitative Livelihood Assessment

Incorporated resources	*Allocative resources*
Work force	Productive goods
Education	Material goods
Health	Subsistence goods
Socially institutionalized resources	*Monetary resources*
International networks	Amount of monetary income
Claims to state resources	Stability of monetary income
Local kinship networks	Diversity of monetary income

Source: Based on Gertel (1997). Adaptations: Ingo Breuer.

Finally, the concept includes monetary resources, a term which refers to three forms of command over money, namely: (1) Amount of monetary income, measured as income per capita and year; (2) Income stability, or the presence of at least one stable income; and (3) Income diversity, which is the number of different economic sectors from which the household's monetary income is derived, and which can be seen as an indicator for the risk-exposure of the household's cash inflow.

Using this concept and these variables, the resource status of the pastoral households of Oussikis was assessed, and inequalities between them were analyzed. This permitted a distinction between those who are highly vulnerable and those who are less so. In the following, the results of this analysis will be presented.

A Resource Analysis of the Ait Oussikis Pastoralists

The analysis revealed that pastoral households in Oussikis show great differences in livelihood security. Most notably, this is reflected in extreme inequalities concerning resource access, but also in marked differences with respect to family management and the role of the pastoral activity within household economies. The following discussion is based on the typology developed above, where pastoral households were classified into three types, namely: Type A as those specialized on pastoral production, type B as those combining mobile livestock production with wage labor, and type C representing those with access to formal sector employment or international migration (cf. Table 11-2).

The households of type A are, most strikingly, much less endowed with livelihood resources than those of the two other types (cf. Table 11-4). Their per capita income is comparatively low, and about one quarter of them live below the poverty line. Concerning the indicators of material wealth, they must be counted among the least well-off in Oussikis. Most of them are nuclear families, and corresponding-

Table 11-4: Pastoral Livelihoods – Resource Access in *Irahhaln* Households

		Pastoral households*			Non-pastoral households
	Unit	Type A	Type B	Type C	
Household structure					
Household size	Pers.	8.5	11.7	16.4	7.3
Adult men in household	No.	1.5	3.9	4.9	2.3
Nuclear families	%	80.0	20.7	7.1	51.5
Extended families	%	20.0	79.3	92.9	48.5
Incorporated Resources					
Labor force	ADR	1.6	0.7	1.1	1.1
Secondary education	%	0	42.1	86.6	26.6
Health (no chronic illness)	%	90.0	78.9	--	87.1
Socially Institutionalized Res.					
Emigrant transfers	%	10.5	31.6	66.7	28.2
State transfers	%	0	15.8	--	25.8
Local networks	%	29.0	21.1	--	21.0
Allocative Resources					
Productive goods: Fields	No.	5.6	5.2	6.7	4.6
Apple Trees	No.	78	90	93	102
Herd	Head	180	148	236	0
Cows/Mules	No.	0.9	1.4	1.6	0.8
Material property	Index	2.9	4.6	4.9	4.7
Subsistence cereal prod.	Kg/P/y	67	45	71	75
Monetary Resources					
Amount of income	Dh/P/y	2629	4716	--	2886
Access to stable income	%	0	26.4	86.6	26.6
Income diversity	No. of IS	--	--	--	2.0
Below the poverty line	%	24.0	9.5	--	35.5

Source: Own survey (n= 170).
Remarks: *defined as >=1 male adult nomad in household; -- = no data; ADR = Age Dependency Ratio; IS = Income Sources

ly, in many of these households a high number of children and seniors depend on the work force of only one male adult (high ADR). None of these households disposes of members with secondary education, which means that they are constrained from entering the more secure segments of the labor market. Moreover, very few of them can rely on international networks. Livelihoods of these pastoralists can thus be regarded as comparatively vulnerable and insecure.

Further analysis, however, revealed that this group can be subdivided into two subgroups. One of them is extremely poor and vulnerable, endowed with very few assets of every category. Concerning their herd management, they show very little herd mobility, which is in many cases due to lack of work force. They thus depend to an extreme degree on local climatic conditions. In the second subgroup, the degree of livelihood security is, despite specialization on pastoral activities, fairly high. These pastoralists are wealthier and better endowed with allocative assets than those of the first subgroup. They also have a more favorable household structure, reflected in a higher number of persons able to contribute to fulfilling the everyday herding tasks. Pastoral mobility is very high in this subgroup, showing flexible adaptation to the herd's needs and to local climatic constraints.

Household type B represents those who combine mobile livestock production with wage labor. The data show that, unlike type A, this is a group with relatively high livelihood security (Table 11-4): They reach a comparatively high per capita monetary income, and only few of them live below the poverty line. This may be due to their ability to regularly transfer work force and money between their different economic activities. As exemplified in the case study, they may, for example, invest in animals in years in which climatic conditions are favorable, while engaging in wage work in years of drought. These households also dispose of a large portfolio of non-monetary resources. For example, they are well endowed with material property and productive assets. About one third of them can additionally rely on emigrant networks. Furthermore, many of these households display low ADRs, which indicates that their age structure further enhances livelihood security. This is mainly due to the fact that many of them are extended families, often with several brothers living together in one household. Summed up, this household type can be addressed as a group of low vulnerability and high livelihood security.

The data seem to indicate that diversification brings about higher livelihood security. However, specialization does not automatically induce high vulnerability. The success of specialization is obviously connected to a flexible herd management, which is in turn dependent on work force availability.

Those *irahhaln*-households with access to formal sector employment or international migration (household type C) are generally rich households, well endowed with productive and material assets. They are well-educated, and often, household members are active in at least four different economic sectors. A particularly important fact is that they have stable incomes which often come from state employment or entrepreneurship. These households are extraordinarily large. This is due to the fact that many households in this group are composed of several brothers who, after the death of their father, deliberately chose to stay together in one household instead of splitting up into several ones. Their economy is quite different from that of the other groups, mainly because they depend predominantly on the transfers of their migrant and/or state employed members. Those household members who are located in Oussikis very often pursue activities which depend on investments coming from the migrant remittances. This is particularly true for their herd economy. Very often, the pastoral activity seemed to be a sort of 'annex' of their other activities, for example, when they used the herd as a 'bank' to stock money gained in the cities. In many cases, they pursued market-oriented ways of animal production, adapting animal sales to fluctuations in market prices rather than to the household's everyday need for money.

Irahhaln n-lkamion – Pastoral Mobility in Change

Traditionally, the Ait Oussikis *irahhaln* followed a pattern with summer pastures in the High Atlas, and winter pastures in the Jebel Saghro and the Pre-Saharan plains. Before the pacification in the 1930s, pastoral mobility was a collective issue. They moved and camped in groups, and movements were organized collectively on the tribal level.

Soon after the pacification, since insecurity was no longer a cause for fear, strategies of pastoral mobility individualized. This has been the case until now. Although there is still a large degree of cooperation among nomads, in the end everyone makes his own mobility decisions, depending on his abilities and the resources at disposal. Thus, today we have a multitude of individual mobility patterns.

Recently, we can see the development of a new form of pastoral mobility, the so-called *irahhaln n-lkamion*, literally: the nomads with the trucks. In fact, some nomads have begun to transport their animals to pastures far beyond the traditional Ait Atta area. To do this, they rent trucks to transport animals and tents. The new pastures may be located as far away as 700 km from Oussikis, and they might change from year to year, depending on regional climatic conditions (Map 11-2). For example, in winter 2001 some Ait Oussikis nomads stayed in several regions in eastern and northeastern Morocco (regions of Missour, Talsinnt, Guercif and Nador). In winter 2002 they went to several regions neighboring the Souss valley and the Anti-Atlas (mainly in the province of Tiznit). Choice of the regions usually demands intense effort, including prospection journeys. In the host region, relations to the local population may be very complex, and they might include cooperation as well as conflict, for example in pasture access and water issues (cf. Werner, Turner, this vol.).

Among the *irahhaln n-Ait Oussikis*, three groups could be distinguished in 2002/03, according to a simplifying typification of their herd mobility patterns (Table 11-5): About one third of the *irahhaln* used truck transport and had transported their herds 'out of the zone'. Generally, these were very large herds, and often had two male adults accompanying the herd. One other third pursued the traditional pattern

Table 11-5: Mobility Patterns and Herd Management of the Ait Oussikis Nomads

Mobility pattern	Herds in Oussikis according to mobility pattern	Distance of displacements	Herd size (head)	No. of adult men accompanying the herd
Ia: Stationary	3.7%	herd: low tents: none	*65	0.4
Ib: Semi-stationary (year-round in traditional summer pasture areas, but still mobile)	28.0%	low (5–25 km)	158	1.1
II: Transhumance between summer and winter pastures (traditional pattern)	34.2%	high (60–120 km)	217	1.7
III: Use of new pastures; truck transport (*irahhaln n-lkamion*)	34.2%	very high (450–700 km)	275	2.0

Source: Own survey (n= 82 herds).
Remarks: *Number of cases too low to create a mean.

of seasonal change between the Ait Atta summer and winter pastures. These *irahhaln* generally had not used trucks, and they own herds of largely varying size. Another third of the *irahhaln* pursued stationary or semi-stationary livestock breeding. Compared to the other types, they displayed lesser degrees of mobility, staying all year long within a 15 km radius of Oussikis. Most of them, however, were not totally settled, but regularly changed camp site in the area around Oussikis. These pastoralists possess relatively small herds, and in some cases herding tasks are accomplished exclusively by children and women.

We are thus witnessing a large differentiation, and even a polarization in pastoral mobility patterns. While some pastoralists have adopted strategies of immobilization, others have expanded their territorial scope, sometimes to an extreme extent. It is, however, important to understand that many of them frequently change their mobility pattern according to climatic conditions. One pastoralist, for example, may choose truck transport in one year, but stay semi-stationary during the next. The crucial question is then: Who is able to choose between different mobility options, and who is not? Resource analysis revealed that several socio-economic factors determine the kind of mobility a pastoral household may pursue. For exam-

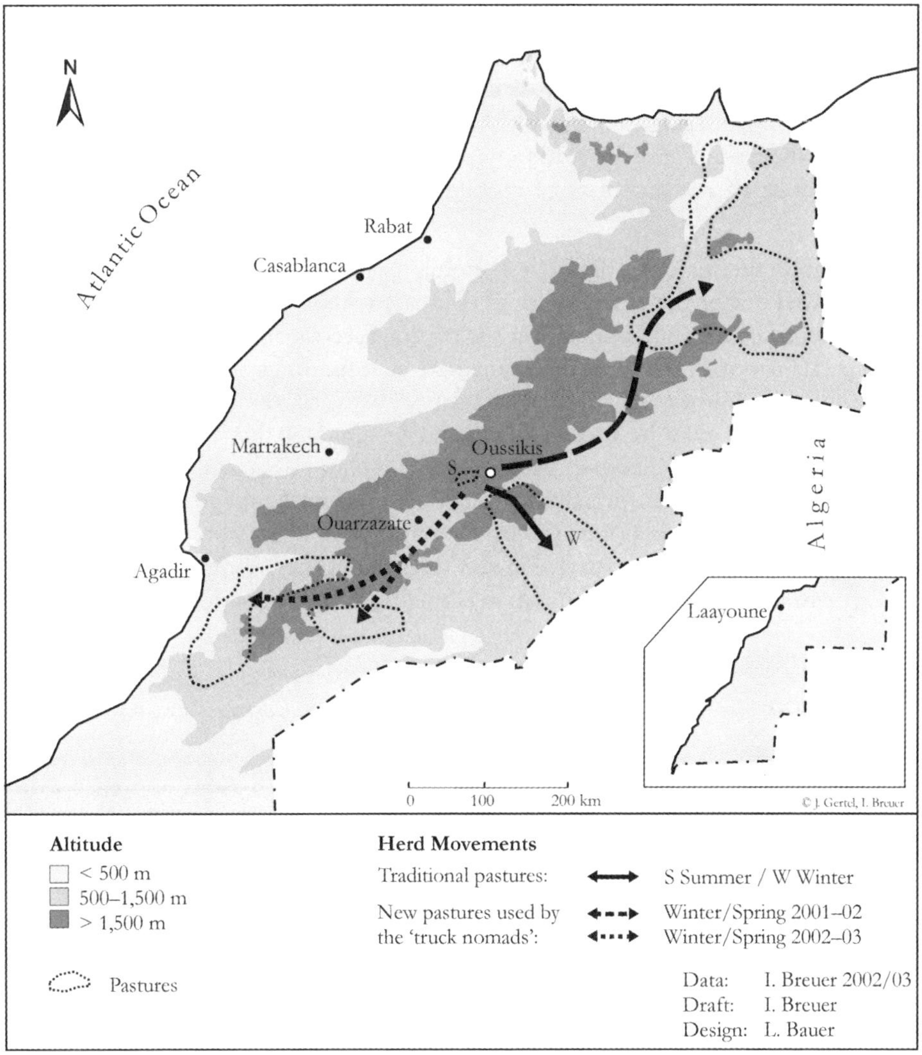

Map 11-2: Pastures Used by Oussikis Pastoralists (2001-2003)

ple, a minimum herd size is needed without which truck use would not be profitable. This threshold may be situated, depending on the year, at roughly 170–200 head. Simultaneously, pastoral mobility depends on the availability of male adult work force. A stationary herd, for example, can easily be kept by only one man and his wife and children, or even by children alone. Concerning the movement out of zone, the pastoralists claim that at least two men are required. This is mostly for (subjective or objective) security reasons, expressed as, for example, 'to protect the herd against thieves' or 'to avoid leaving the women alone in the tent' in a social environment that is, to many of the pastoralists, partly unknown. Out-of-zone-mobility is thus only open to those who manage to maintain a household structure allowing for sufficient male work force. Another factor is the pastoralist's social skills in gathering information on and organizing access to remote pastures. Although these skills are crucial for any pastoralist to successfully conduct the herd, in the context of the out-of-zone-mobility, they have gained heightened importance.

Accordingly, the majority of the less mobile pastoralists (groups Ia and Ib) have never moved out of the zone. Many of them are compelled to stay, constrained by their weak resource endowment. On the contrary, most of the *irahhaln n-lkamion* (group III) choose to stay in the immediate surroundings of Oussikis whenever pasture quality allows. The close interrelation between resource endowment and pastoral mobility may be further illustrated by the fact that two of the pastoral household types defined above (Tables 11-2, 11-4) have a particularly large share of *irahhaln n-lkamion*: the 'specialized-mobile' ones (type A, subgroup two), and the 'rich-and-diversified' ones (type C). In the first case, new mobility may be regarded as a necessary condition to survive in the absence of other resources. In the second case, pastoral mobility clearly aims at maximizing herd production.

Conclusion

In Oussikis, like in many other communities of the High Atlas and even other Moroccan regions, pastoral activities are deeply embedded in broader livelihood strategies of multi-activity, multi-local households. In this context, herds may serve very different economic functions, according to the household's wider asset and activity portfolio. Strategies like maintaining a favorable household structure, or diversifying economic activities, are not only crucial for reducing vulnerability, they also have a considerable impact on how herds are conducted, and thus how pastoral strategies are constantly re-modeled. Future research into pastoral systems should take careful account of economic multi-activity in pluri-local household contexts.

Notes

(1) Fieldwork comprised both qualitative and quantitative methods. To gain insight into qualitative aspects of the local livelihood systems, I engaged in participant and non-participant observation and semi-structured interviews, as well as long-term observation of daily life in a small number of households. In a second step, I conducted a micro-census in all of the households in the study area (n= 373), in order to gain

information about each household's demographic structure, as well as economic activities and mobility profiles of household members. I also conducted standardized in-depth interviews with a stratified sample of 203 households. These interviews focused on resource access and use, and they provide the data basis for the resource analysis.

–12–

Pastoral Livelihood Strategies in the Draa and the Souss

Jutta Werner

Extensive mobile pastoralism[1] is the only form to use the common pastures in an ecological adapted way in the arid south of Morocco. Other forms of livestock keeping or even agricultural production are economically and ecologically unsustainable in this area, as they require expensive irrigation or additional fodder import. However, mobile pastoralists in Morocco and elsewhere[2] face various constraints due to dramatic changes within their socio-economic environment. In general, the importance of mobile pastoralism is not acknowledged in the national, as well as in international donor policies. Stationary agricultural production is rather favored, even on less productive sites, and land use models that are aiming for private property and individual land rights are promoted. The poor organizational level of mobile pastoralists in professional associations contributes to this. Consequently, the present situation of mobile pastoralists in southern Morocco can be characterized by five factors: a restricted mobility due to the strong increase of agricultural land use; a low standard of living, since medical supply and schooling possibilities are largely missing; an increasing number of conflicts between different groups of resource users related to access to collective land; an insecure and ambiguous legislative basis for land use rights; and few and non-adapted public investments in the area.

How do mobile pastoralists act in this situation? What strategies do they apply in order to obtain livelihood security in a sustainable way and which factors determine the success or failure of pastoralists' strategies? How can it be explained that some mobile pastoralists become increasingly oriented towards entrepreneurship, while other pastoralist groups impoverish? In order to find answers to these questions this paper discusses the livelihood strategies practiced by south Moroccan pastoralists, their reactions and responses to risk factors, including governmental or international donor policies. The case studies are located in the Draa valley in southeastern Morocco and in the southwestern Souss region.

Table 12-1: Movement Pattern of one Pastoralist Based in Tiznit (1994–2003)

1994	1995	1996	1997	1998	1999	2000	2001	2002	2003
Tiznit	Tiznit ↓ 300 km Tansift ↓ 500 km Tissint ↓ 392 km Tiznit	Tiznit ↓ 261 km Chichaoua ↓ 160 km Amskroud ↓ 148 km Ifni ↓ 50 km Tiznit	Tiznit	Tiznit	Tiznit ↓ 842 km Gueltat-Zemmour ↓ 842 km Tiznit	Tiznit ↓ 350 km Abda ↓ 84 km Rehamna ↓ 123 km Khouadra ↓ 147 km Essaouira ↓ 280 km Tiznit	Tiznit ↓ 1,020 km Guercif ↓ 267 km Sidi Qacem ↓ 82 km Fes ↓ 10 km Sidi-Harazem	Sidi-Harazem ↓ 175 km Guercif ↓ 140 km Nador ↓ 1,160 km Tiznit	Tiznit ↓ 341 km Haouz

Source: Own survey

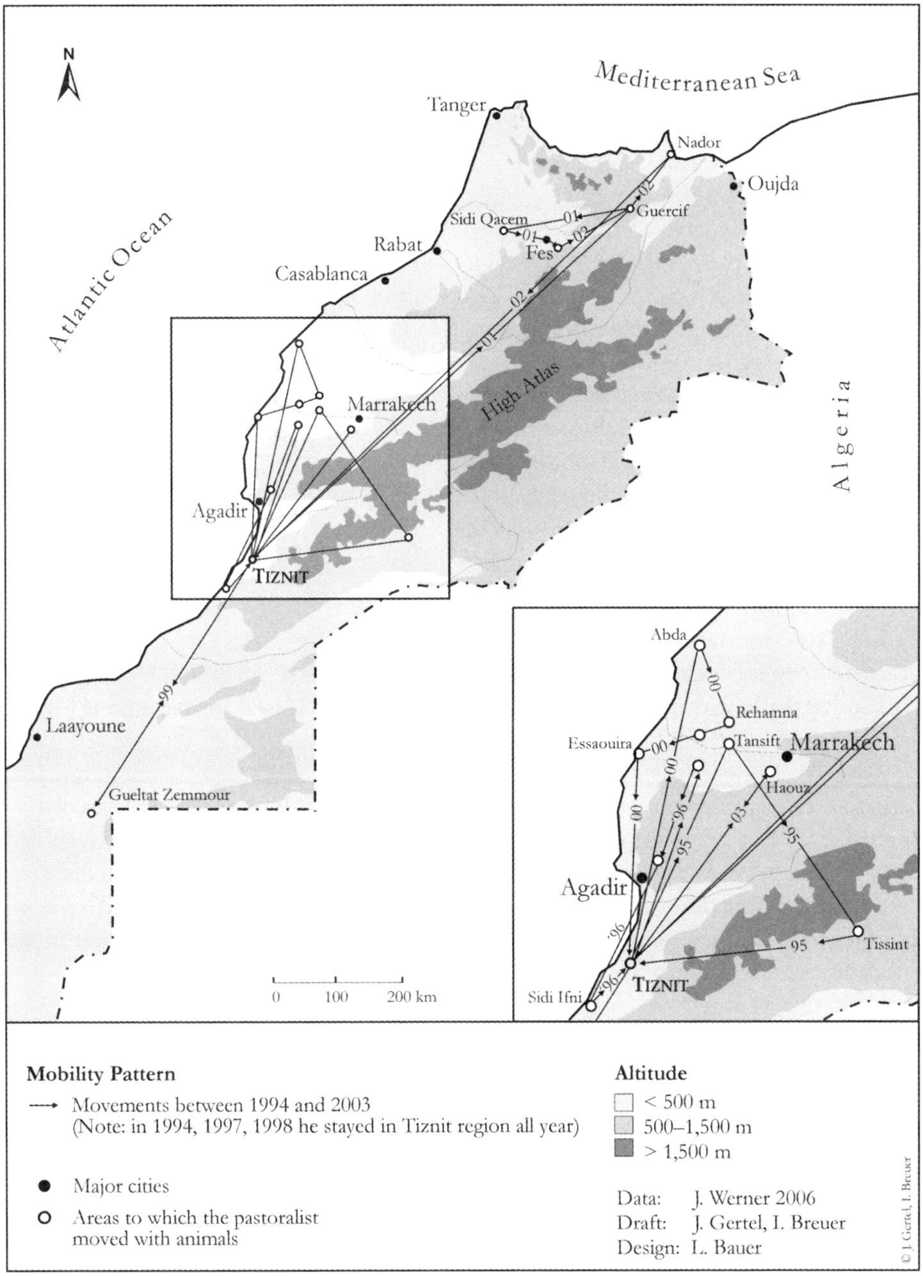

Map 12-1: Movement Pattern of one Pastoralist Based in Tiznit (1994–2003)

How to Cope with Risks?

Generally, nomads in the region employ two main strategies to manage upheavals in their livelihood setting: some specialize in mobile livestock production and others diversify their income sources (Fig. 12-1). Households from the Draa valley that are specializing in nomadism cover great distances. During droughts they move, for example, to the semi-arid Souss region by using trucks, covering a distance of between 500 and 700 kilometers. Others, like the Souss based pastoral household represented in Map 12-1, have been moving to various, widespread regions throughout the country during the last 10 years, with an extremely high frequency (cf. Table 12-1). However, this coping strategy is limited, as the Souss example reveals: During the last severe drought in southeastern Morocco (1998–2003) nomads from other southeastern regions also migrated with their flocks and installed themselves in the Souss region. Since Sahraouis, originating from the Sahara provinces, are also pasturing their herds in the Souss, increasing pressure on the common resources fosters violent conflicts between the autochthonous sedentary population and the pastoralists (Turner, this vol.).

For the pastoral households in the Draa valley aiming to diversify their income sources, irrigation farming on former collective lands is the preferred strategy. Other strategies are to join the professional army, to engage in smuggling[3] or to become active in the tourism sector (Fig. 12-1). These sources of income subsidize pastoral livelihoods particularly during droughts, while in rainy years nomadic pastoralism remains the most profitable economic activity. Generally, the engagement for additional income requires a permanent presence in a certain place and thus limits pastoralists' mobility. Consequently, large quantities of supplementary feeds and concentrates are required to feed the herds during droughts. This contributes to a situation where the herds are kept in a quasi immobile manner on the pastures, while fodder is bought in towns and transported to the herds. The limitations of this strategy are obvious. The use of pastoral lands as livestock 'parking lots' (Blench 2000, 26) fosters an increasing pressure on ecological resources, causes degradation and renders the pastoralist even more vulnerable.

Pastoral Livelihood Strategies: Three Case Studies

Given the complexity of livelihood strategies, three case studies will be discussed in more detail.[4] While these households pursue different strategies, each seeks to minimize risks, to increase income, security and the well-being of the family. The first household engages in additional cropping activities, the second disposes of two permanent sources of income, and the third specializes in nomadic pastoralism as a livelihood strategy.

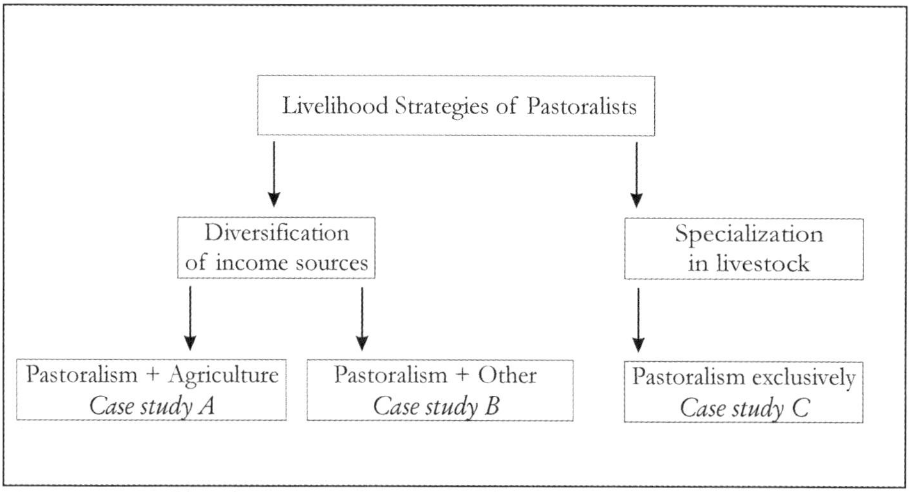

Figure 12-1: Livelihood Strategies of Pastoralists from the Draa Area (source: author)

Case Study A: Diversification – Pastoralism and Agriculture

The first household represents the majority of nomadic households in the Draa valley. Like almost all nomadic households, it employs an income diversification strategy through irrigated agriculture on former collective land. Indeed, all households of the sample (n= 82) considered cropping as the most successful strategy to secure their livelihood during drought. One pastoralist owning a farm in the Draa valley expressed it this way: "We can no longer imagine a life without the farm. During droughts, God and the farm help us a lot."

Household members:	*Parents, 9 children (between 9 and 25 years)*
Ethnic group:	*Ait Imsouffa (origin and home in the Draa valley)*
Livestock:	*90 goats (78 females, 12 males)*
	30 sheep (during drought years permanently on the farm, during rainy years on the pastures near the tent)
	11 dromedaries (10 female, 1 male; including four belonging to the pastoralist's brother)
	Donkeys
Farm:	*Size: 2 ha (only part of this is cultivated)* *Status of the property: melk*

Since 1995 the household is divided between the tent and the farm. One son and his own family stay on the farm to ensure the daily farm work, while his parents and some of his siblings live in the tent to take care of the herd. For seasonal work, like sewing and harvesting, some of the children or the mother may also help out on the farm. One 8 year old son stays on the farm to visit the nearby school.

Apart from crops destined for self-consumption, the household cultivates henna as a cash crop. During the years of drought (1998–2001) the henna price was very high (17.50 Dh per kg) compared to the 1980s when the price of one kilo of henna was at about three dirham. By selling henna the household could ensure its livelihood during the drought, and was even able to buy concentrates and other forage for its livestock during the entire period. Accordingly the animal mortality was not higher than in rainy years. Indeed, they fattened some sheep and sold them in order to buy animal fodder. They also kept some small he-goats and sold them only after the drought when the prices had increased again.

The difference between dry and rainy years concerning the household's resource management is that the money used to buy fodder for the animals in drought years can be saved during rainy years. Following the information obtained from the pastoralist an amount of about 20,000 Dh is used to buy fodder during a year of drought. Due to saving this amount in rainy years, this household has been one of the few with a high number of lambs (33) born in 2002, an especially dry year.

The members of the household do not, however, appreciate the farm work. They say that once they are getting old, they will still live in their tent and not move to the farm. But according to their estimations, mobile pastoralism would neither be profitable nor possible without the extra income provided from the farm.

Since this household runs a farm with access to pump irrigation, it has a relatively secure income during droughts and thus – in contrast to other nomadic households – is not obliged to sell large parts of its herd in order to buy fodder for the remaining animals. Due to this, they do not need to keep a large herd. They are able to sell their cash crops, like henna, at high prices even during droughts, which allows to cross-subsidize animal production. Hence, agricultural production is buffering and spreading risks inherent to pastoral nomadism. Albeit this agricultural engagement does not represent an ultimate decision for sedentarization. While agricultural and pastoral production are complementary to each other, pastoralism remains the most profitable income during rainy years, and even allegedly 'completely' sedentary households remove to the tent in extraordinarily rainy years. Sedentarization thus remains reversible (cf. Galaty 2004a, 2).

Case Study B: Diversification – Pastoralism, Tourism and Army Employment

The second case concerns a household with two permanent income sources: from employment in the army and from a tourism agency using the pastoralists' dromedaries for tourist desert excursions. This household lives distant from traditional forms of nomadism. They do not own a tent any more and the pastoralist qualifies himself as an entrepreneur in the tourism sector rather than defining himself as a nomad. He uses a part or even his entire herd for tourist desert excursions. This form of engagement in tourism is only accessible for few pastoralists – those who can read and write. Telephone and fax are required. As the pastoralist is obliged to travel frequently to acquire new clients and also has a fixed office in town, he must employ herders to care for his stock. The herd thus stays at least partially mobile.

Household members:	*A man with his parents, his wife and two children (3 and 5 years old), the whole family lives in a house in town* *Two employed herders*
Ethnic group:	*Aarib (originally from the Sahara provinces, now at home in the Draa valley)*
Livestock:	*80 dromedaries (30 for tourism, 50 for breeding)*

The household employs two herders, one of them for the dromedaries used for the tourism business, the other for the dromedaries held for reproduction. In addition, the household disposes of a traditional house, a concrete house and date palm fields in the Draa oasis. In addition to the fluctuating income from mobile pastoralism and tourism, there is a permanent income from the army, where the pastoralist is employed as a corporal.

The 50 dromedaries kept for reproduction only contribute to the household's revenue in rainy years. They then furnish one third of the overall income. This is why the dromedaries are not abandoned completely, although the pastoralist's revenue does not significantly depend on them. He considers the dromedaries kept for reproduction as a "relic of his parents' nomadic times". If he cannot find a herder he leaves the dromedaries alone in the desert, without knowing exactly where they are. During droughts they receive no supplementary feed – in contrast to those kept for tourism. Consequently, their number diminished within the last years from 110 to only 50 animals. The dromedaries used for tourism, however, are very well kept. During the tourist season they stay in a paddock in town. They are mobile only in summer time, the period without any tourism, and are kept by a salaried herder.

There are significant differences in the income composition between dry and rainy years. In dry years two thirds of the household's revenues (about 6,000 Dh) are created by tourism activities and one third comes from employment in the army (about 3,000 Dh). Furthermore during drought, two thirds of the household's total income is used to buy fodder and veterinary products for the dromedaries. This money represents the benefit during rainy years. In rainy years, one third of the total household income is generated by selling dromedaries (about 6,000 Dh), and one third by tourism (about 6,000 Dh). The income from the army (about 3,000 Dh) and the income from the date palm trees (about 3,000 Dh) contribute each one sixth of the household's total income. The income from the army is entirely devoted to the pastoralist's parents.

Due to its permanent revenues, the household does not depend on mobile pastoralism and consequently does not encounter any risk during droughts. It is a typical example of a household having abandoned mobility and life in the tent for the sake of a profitable life in town with a permanent non-pastoralist salary.

Case Study C: Specialization – Pastoralism

The third household specializes in nomadism and moves frequently.[5] They are obliged to move to very distant, isolated sites during droughts. In southern Morocco the specialization in nomadism can only be profitable with a large herd size. Research indicates that these households normally dispose of at least 200 sheep or goats (cf. Breuer, this vol., Pre-Sahara). Sometimes such households split during severe droughts. The sheep and goats then move to the Souss region, while the more drought-tolerant dromedaries rest in the arid southeastern regions.

Household members:	*Parents, 1 daughter-in-law, 7 children (between 1 and 26 years)*
Ethnic group:	*Ait Khabbach (originally from the Ziz valley, now at home in the Draa valley)*
Livestock:	*200 goats*

This middle-sized household specializes in mobile pastoralism as the family business, without (significant) other income sources. The household has its home in the Draa valley, but, because of the drought in the Draa valley in 2000, moved to the less dry Souss region. Thanks to the household's movement, the pastoralist could prevent a major herd decrease and did not need to buy additional feed or concentrates. After abundant rainfall in the Draa valley, the household moved back again during spring 2004.

Despite severe problems – including serious difficulties in getting access to the pastures and very complicated relations with the autochthonous sedentary population – the household does not regret its move to the Souss. The pastoralist appreciates especially the fact that he was not obliged to buy supplementary feed.

Concerning mobile pastoralism as professional activity, the two generations in the household (parents and children) have conflicting opinions: the father appreciates pastoral nomadism very much, but the sons strongly dislike it. They do not like to help out with activities related to nomadism, but are rather looking for work outside the pastoral economy, for example on the surrounding farms. For this reason smaller children (aged 5 and 6) are responsible for the herd. They know much less about herding techniques and cannot go to places too far away from the tent. As a result of this conflict, the eldest sons did not accompany their family back to the Draa valley, but preferred to stay in the Souss region, because of its economic advantages. The contribution of the sons' labor during drought years on the farms in the Souss region is at about 3,000 Dh for about one to three months per year.

In rainy years, the household earns about 22,000 Dh from goat sales. They have invested these benefits in a construction lot in the Draa valley. This purchase represents a risk minimization and diversification strategy and is further thought to ensure the livelihood after the nomad's retirement.

Mobile pastoralism in this household is profitable due to the high moving frequency to distant and isolated places. The herd size has been stable during the last ten years, and profits were not invested in animals, but rather in property. The sons' employment as seasonal laborers on the surrounding farms and the purchase of a construction lot indicate that the household starts to diversify its income sources.

Adapting versus Coping – How to Determine Success or Failure of Pastoral Strategies?

A comparison of these three examples underlines that significant differences exist not only in the types of livelihood strategy employed, but also in their efficiency and success. The result is a disparate and specific vulnerability of pastoral households.

The three case studies also reveal that the specialization strategy is much less successful at reducing the household's vulnerability than the diversification strategy. This is further emphasized by a differentiation between adapting and coping strategies.[6]

Adapting strategies are considered to be short-term reactions to external shocks such as droughts or conflicts. Coping strategies, in contrast, are comprehended as long-term strategies seeking to improve the household's situation by minimizing its vulnerability. Collecting and selling meteorite pieces to tourists during droughts, for example, is a short-term reaction to a scarcity of liquid financial resources and therefore an adapting strategy. The effort to enroll a second household member in the professional army in order to have a second permanent salary apart from pastoralism represents a long-term strategy and is therefore called a coping strategy. However, the boundaries between these categories can be fluid. Some strategies can either be seen as adaptive or as coping, depending on how they are employed. Smuggling, for instance, can be an adapting strategy if the household smuggles only in cases of financial need during droughts, but for a pastoralist regularly engaging in smuggle, it is a coping strategy. An additional income source in tourism can be an adapting strategy for a pastoralist who occasionally rents his dromedaries to a tourist agency, but who cannot calculate on regular incomes to minimize his vulnerability. For a pastoralist possessing his own tourism agency, however, the tourist sector represents a coping strategy oriented towards long-term revenue.

It is possible that what starts as an adapting strategy becomes a coping strategy. An example is a pastoralist working as a day laborer during droughts who finds that little by little this becomes a regular income source or maybe even the only one. This case is not very frequent because most pastoralists do not have access to permanent income sources apart from pastoralism and agriculture. This lack of access to alternate employment is mainly due to the fact that nomadic pastoralists often cannot read and write. Additionally, they do not have the financial resources necessary to access permanent employment, such as the army offers.

The contrary is also possible. A strategy practiced by a pastoralist as a coping strategy can lose its characteristics. Due to changes to the socio-economic framework, the vulnerability of a nomadic household can increase considerably. This means that the strategy can no longer ensure the nomadic livelihood and must therefore be considered as an adapting strategy.

Households in which specialization is a coping strategy have an accentuated vulnerability. Presently, the specialization in nomadic pastoralism can still be seen as a coping strategy. Yet, considering that the possibilities of access to good rangeland will further deteriorate, it must be questioned if this strategy will still represent a coping strategy in future, especially since violent conflicts related to the access to pastoral land are becoming more and more frequent (cf. Turner, this vol.). The households, which are currently specializing in nomadic pastoralism, risk running out of time to find alternative coping strategies.

Long-term coping strategies are, in comparison to other livelihood options, most promising for economic success. Households with permanent income sources apart from pastoralism are those with the highest standards of living and are still improving these standards. These households are also the least vulnerable during droughts, for the following reasons: For one, these pastoralists are not forced to sell a part of the herd during droughts to ensure the survival of the other animals. Because of the additional income source, they are able to feed their herd with supplemental feeds and concentrates. By securing the herds during droughts, these households can profit from livestock sales in rainy years, and are not forced to move to isolated and distant regions in search of pasture. However, if the pastoralist wants to move, he will probably have cash in order to pay the truck transport of the herd, as well as to pay some 'fees' to gain access to fertile pastures.

The engagement in additional income sources, like agriculture or tourism, however restricts pastoralists' mobility. Salaried jobs limit mobility to a much lesser extent, because generally only one or two members of the household are concerned, while the other members are able to move with the herd. In any event, reduced herd mobility leads to increased pressure on the ecological surrounding, which will be discussed in the following.

Ecological Sustainability of Pastoral Livelihood Strategies: A Dilemma

The economically most sustainable strategy is not necessarily the most sustainable strategy from an ecological perspective. Ecologically, the specialization in nomadic pastoralism, with a high moving frequency and the use of isolated rangelands, both of which limit overgrazing of the pastures, are sustainable. The respective households are obliged to adapt their herd size to the capacity of fodder the pastures provide by selling animals.

The strategy of income diversification through revenues from agriculture or from a permanent employment (army) does not contribute to a sustainable management of the common natural resources. Often, groundwater is not sufficient for long-term irrigation farming. Several farms in the arid Draa valley have already been deserted for this reason. Cropping also requires the pastoralist's presence. Frequently, herd owners do not want to leave their livestock alone with an employed herder in an isolated region. Hence, they limit the moving radius of their herd to a few kilometers only, leading to a high pressure on the pasture around the farm. Pastoralists further tend to keep their herds near urban centers and roads in order to facilitate fodder transport, and it is this group that does not adapt their herd size to the fodder available on the pastures. In case of drought, these pastoralists neither move to other places, nor reduce the number of animals to which they transport fodder. In sum, this leads to an overgrazing of the pastures near the farms and causes degradation.

Adapting strategies – irregular income strategies aiming to alleviate a crisis – practiced by impoverished nomadic pastoralists, can limit mobility, too. An impoverished nomadic pastoralist engaging in an irregular additional income – for example by renting his dromedaries to tourist agencies – cannot move out of a drought prone region. As a livestock producer, he is rather looking to install his herd near an urban center. Being impoverished, he does not possess the power necessary to install his herd on a better pasture. Very often, this results in overgrazing. However, as impoverished pastoralists are forced to sell animals during droughts to ensure the survival of household members, their herd size diminishes and so too does the pressure on the pastures.

In conclusion it can be said that a herder aiming to reduce his economic vulnerability cannot practice an ecologically sustainable form of pastoralism. Under the given conditions in southern Morocco there is no production system reducing pastoralists' vulnerability and allowing at the same time an ecologically sustainable pastoral production. Vulnerability reduction in an economically sustainable way means income diversification, and thus a strategy increasing the pressure on the pastures. For this reason nomadic pastoralism is under a real dilemma.

Notes

(1) In the context of this paper 'mobile pastoralists' are seen as those people who employ some form of transhumance or nomadism in their extensive livestock production system (see UNCCD 2002, 1). The concerned groups designate themselves in most cases as 'nomads'.

(2) See for example: Abbas (2004) for the case of Algeria; Galaty (2004a, b) for the case of north Kenya; Janzen (2004) for the case of Mongolia; Gertel (2002); Rao (2002) for Asia; Streck (2002) for the case of Sudan; Niamir-Fuller (1999) for the case of Uganda; Ruf (1995) for the case of Mauritania; Abdullahi (1990) for the case of Somalia.

(3) These income sources are specific for the Draa valley. In other regions like the High Atlas other income sources, like working in the construction sector or international migration, are also employed (Breuer, this vol., High Atlas).

(4) The case studies are selected from 82 studied households (Werner 2006).

(5) Households specialized in nomadism, without income sources apart from pastoralism, are less frequent than households with diversified income sources.

(6) See DFID (2001, section 2/30); Fläming et al. (2001, 20 ff); Goldman et al. (2000, 1–2); Scoones (1998, 9); Werner (2005, 156 ff).

–13–

Social Lines of Conflict between Pastoralism and Agriculture in the Souss

Bertram Turner

The Souss in southwest Morocco has historically been an area of interaction between a local resident rural population and mobile pastoralists, a situation which continues to date. The Argan forests, which form a worldwide unique ecosystem, are an essential part of the livelihood of both nomads (rahhalin) from various regions practicing different types of mobile breeding on the one hand, and local farmers and peasants (fellahin) on the other hand, whose activities vary from conventional forms of rain-fed cultivation and irrigated agriculture to hypermodern agrarian production.

However, the established social relationships between both groups of actors have experienced tremendous changes in the course of the last several decades. The subsequent increase of violent conflicts between both groups has been ascribed to specific transformations. These transformations may be analyzed in a comparative perspective as a shift from traditional relations of complementarity and cooperation in resource exploitation to relations of confrontation due to accelerated competition over access to increasingly scarce and therefore contested resources. The changes in many pastoralist-agriculturalist relationships have often been described in terms of conflict generation by the substitution of complementarity through competition and confrontation.[1]

I argue here, that this model does not fully apply to the situation in the Souss, as mono-causal explanations always prove to be insufficient. Rather, changing interests, shaped by transnational and external impulses of a global and neo-liberal economy, drive a transformation of established relationships from traditional complementarity of the ideal type – including the connotation of a sustainable and ecological balance between various nomadic and various local groups – towards a complementarity between strategic groups[2] within the post-protectorate elite.[3] This elite developed during the transformative process itself. The concept of strategic groups also suggests that in the course of transformation, actors form new networks in order to secure privileged access to previously underexploited resources. Horizontal networks of cooperation are combined with vertical cushioning of established power structures. The paper thus argues that analysis of the wider social context reveals that there is no transformation from complementarity to conflict at all, but a change of constellations of cooperating and/or conflicting actors. Horizontal alliances on a high level of social rank developed and delegated the conflict potential resulting from intensified competition to social actors who had no 'exit option'.[4] As a consequence, the 'voice option' which includes reverting to violence came to the fore. In the concrete case of the Souss, long distance camel herding pastoralists are interacting with real estate holders. Central to the analysis is the conceptualization of that relationship within a frame of informal local order. It is sometimes called '*code pastoral*' with respect to the existence of customary regulations that were accepted by both sides in the past, but which the new arrangements rule out.

Three historical developments are crucial in order to evaluate cooperative versus confrontative relations between mobile pastoralists and local farmers. First, after independence agrarian and irrigation development initiated under the French protectorate continued. This development resulted in a concentration of landed property in the hands of a local rural elite.[5] Second, there are the effects of the reintegration of the Moroccan Sahara into the kingdom after the *Marche Verte* in 1975. It was the political will to encourage Sahraoui camel herders to consider the whole of Morocco as their legitimate pasture ground (agdal). These politics had enormous consequences and contradicted many local regulations of shared resource utilization between *fellahin* and *rahhalin* and led to the reduction and even destruction of existing social ties with other formations of mobile herders who were used to staying in the Souss seasonally. Third, there are the effects of transnational intervention in the form of world market economy, development cooperation, human rights and other initiatives, which are reflected in concepts such as sustainability, environmental protection and the politics of decentralization, in the setup of civil societal structures but also free market trade or in transnational Islamic movements. Articulations and historical conjunctions between these processes of very different qualities led to a breakdown of the established and ecologically more or less sustainable interaction between agriculturalism and nomadism. Different actors exercising different modes of production emerged and dominated in the exploitation of increasingly scarce natural resources.[6]

The Local Setting

The Souss is a plain surrounded by the mountains of the Atlas and Anti-Atlas in southwest Morocco. Most of its population, the Souassa, are *fellahin*, farmers and peasants. Local forms of agriculture, combined with small-scale livestock husbandry, are diverse. There are zones of conventional and zones of hyper-modern agrarian production, irrigated zones as well as zones of rain-fed cultivation. Large citrus and vegetable plantations as well as greenhouses dominate the landscape in the central plain. Agrarian activities and irrigation remain important points of transnational connection today. Animal husbandry is an indispensable element of the local rural economy. The stocks are normally composed of sheep and goats. A few cattle may also be raised in the rural household. Livestock plays an important role as a capitalizable reserve and as supplier of meat and manure, of which the latter is a prerequisite for irrigated agriculture.

The Souss is also home to the Argan forest, a unique ecosystem and object of transnational development intervention. The Argan tree (Argania spinosa L.) is an endemic relict of the Tertiary period and forms the only tree building forests in this semi-arid area able to curb progressive desertification. The Argan forest provides the essential means of livelihood for large parts of the local population. However, the multiple forms of exploitation of the forest, in the first place as pasture, have been considered responsible for the continuous degradation of the Argan tree population. The herds of the resident population are by more than two thirds composed of goats. The adapted species climbs the trees and grazes in the crown. The Argan forest is also incorporated into the zone of agricultural activity. Cereals are grown in rain-fed cultivation and irrigated vegetable production started to expand in the Argan forest. The range of products the forest provides includes barley and oil from Argan fruits as staple foods, as well as vegetables produced as cash crops and many additional products such as wood, mold, medicinal plants and what can be obtained from herding. In 1998 the whole area of the *Arganeraie* was classified as a UNESCO Biosphere Reserve. Although the realization of the project remains for the most part theoretical, the resulting consequences for legislation, legal practice and the management of natural resources in general contribute to the central problems addressed in this paper.

The majority of the local Souassa are *Tachelhait*-speaking Berbers, whose self-designation is *Ichilhayen*. However, there is a high proportion of Arabic-speaking people in the plain, concentrated particularly in the urban centers and in some enclaves in the countryside. Clan affiliation, the patri-lineage system and segmentation, as well as the elaborated alliance systems of the Souassa have given rise to extensive debates in anthropology.[7] However, descent criteria do not dominate social relations today. Furthermore, increasing mobility and long-distance marriage arrangements have modified the allocation of descent ties in space in recent decades. In times of crisis however, they may be reactivated and reconfirmed as reference units for collective action. Further social criteria of inclusion and exclusion such as neighborhood solidarity contribute to the local conception of social cohesion. This type of socio-political organization, called *taqbilt* (berb.) or *qabila* (arab.), represents the basic reference frame for local identity. It is a named territorial group that understands itself as a social unit, in which solidarity obligations transcend the descent group. These units are sometimes referred to as tribes, sometimes as tribal fractions and sometimes as completely different types of social units. They persist as the basis for the rural districts and have thus been integrated in the Moroccan state administration. A rural district in the Souss consists of a number of villages and hamlets. No dominant pattern of internal social organization exists in these villages. Some are composed of or dominated by members of one lineage. Others consist of members of a number of quite equal lineages. Political organization in these rural districts oscillates in practice between tendencies of centralization and dissemination of power. This concerns both official and informal institutions. Whether based on informal organization or integrated in state administration, in some villages prominent figures or elitist oligarchies of wealthy landholders dominate the political sphere. In other villages, informal collective organizations take responsibility for leading and organizing the community.

Sahraoui nomads from the Western Sahara with their large camel and goat herds invade the region regularly. During the dry season, nomad encampments are numerous in the Argan forest, even in the vicinity of the villages. In the past sporadic Sahraoui settlements have been installed with which the mobile groups maintain close relationships.

These factors sketch the patterns and initial positions for a more profound evaluation of the social dimensions of conflict. A diachronic perspective should however also be kept in mind. The Souss is steeped in history and state influence has always been significant. For centuries the area has been connected to the outside world through trade and migration. It is also a center of peaceful, scholarly Sufism, represented by venerated and erudite specialists. However, the history of the region has been violent. The local Islamic Sufi movements were involved in military activities and the region was a center for both religious and political movements decisive in the political history of the whole of Morocco.[8] Given this history, recent conflicts between powerful interest groups or alliances as well as external influences from development organizations and religious movements are not something completely unfamiliar to the Souassa. A local peasant explained to me his attitude in a critical situation in reference to a sort of cumulative historical repertoire: "We [!] have chased away the Portuguese [in the 15th century] and we will do the same with our enemies today."[9]

Agriculture and Mobile Pastoralism in the Souss

The empirical data confirm the basic assumption connected with research on nomadism in general, that mobile pastoralism cannot be disconnected from a given local agricultural environment. As has been said, the Souss has been an area of interaction between mobile pastoralists and local resident farmers and peasants for centuries. This implies local arrangements of access. Allied groups practicing transhumance with small livestock came regularly during the dry season when pasture was accessible after harvest, most of them from the region east of the Souss, located between Zagora and the slopes of the eastern Atlas.[10] The draw wells (naora) for irrigated agriculture did not produce a surplus of water at that time. Only the shared use of cisterns (metfiyyas) was possible and this regulated the size of the herds in a particular area. The water stocks of the *metfiyyas* were sufficient for both the herds of the locals and the nomads over the summer period. Accordingly, the exploitation of the forest as pasture was constrained by the limited access to water. This system fed the ecological dream of a complementary relation based on a sustainable balance and adapted to local natural conditions. The physical access to the forest was further restricted by its impenetrable thicket. Thus, only the areas used by the resident population were also accessible for the nomads.

The mostly *Tachelhait*-speaking transhumant partners of the local peasants who were integrated in these arrangements of resource sharing, however, could not maintain their position. For political reasons, one particular group became dominant: the Sahraoui. Already in the past some units of camel-raising Rgueibat, Ait Oussa and other nomads of the south maintained contractual relations with partners in the Souss. They provided animals for caravan transports and camel meat for religious festivals. Some of these traditional contractual arrangements still continue. The upheavals in the 1950s in the Spanish colonized Rio de Oro province motivated some Saharan groups to stay in the Souss, particularly in the buffer zones that traditionally separated areas claimed by local villagers for their respective exclusive use.[11] Still today, these settlements stand out from their environment in appearance and social lifestyle. Another wave of settlement formation took place in the 1970s when the southern provinces were reintegrated. The first wave of pseudo-sedentarization of Sahraoui formations coincided with the acceleration of agricultural development. These circumstances opened better access to the forest for the camel herds. The ultimate shift came with the integration of the region, including the forest, into the world market system. The introduction of new cash crops played the decisive role here.

The integration was enforced during the French protectorate (1912–1956). The modernization of local agriculture started in the mid-1940s with the foundation of 'colons farms' producing citrus fruits for the European market,[12] while local landowners were forced to sell land as a result of precarious living conditions. Simultaneously a significant clearing of the forest areas began. The new farmers dug deep wells and installed motor pumps. With the first expropriations after 1973, the situation changed completely. Moroccan investors had been able to take over large parts of the land as individual private property. In the aftermath, extensive agricultural production resulted in an overexploitation of groundwater reserves. The constant deepening of the wells and the thus increasing costs of fuel for the pumps made water expensive, which resulted in rationalization. The former peasants who had sold their land and worked on the farms as wage laborers lost their income, because the newly introduced irrigation technologies required less manpower. At the same time the introduction of new cash crops, particularly the tomato, allowed them to reactivate traditional sharecropping practices and to participate in modern agricultural production (Turner 2003).

> The tomato became the poor man's citrus fruit. Also we small peasants could from then on profit at least a bit from the 'grand souk' [world market].[13]

Since tomato production demanded uncultivated land, the zone of agricultural activity was expanded into the Argan forest. As a consequence, the rhythm of seasonal production in the forest was redefined. The brushwood has been cleared in large parts of the forest to be replaced by irrigated fields.[14] From then on, these parts of the forest offered open access for camel herds. The nomads came in greater numbers, and not only those with whom the peasants had established contracts. Competition over access to the increasingly exploited forest areas was inevitable. In the course of time and since lasting droughts have continued for several years, the number of encampments and the size of the herds, especially the camel herds, are increasing even more.[15] In the forest surrounding the *commune rurale* considered here, there were four camps in summer 1996 and about 23 in summer 2001. In the whole plain of the Argan forest the number was about 280 camps at that time with 30,000 camels altogether.

There are, on the other side, the large estates, which were never directly affected by the increase of mobile pastoralists since the herds have no access to these farms, as they are protected by impenetrable cypress fences or thorny jujube hedges and often also by walls. These two groups of actors had little contact until they realized their complementary interests and their potential for cooperation. Since the large farms dispose of their own wells and since the introduction of water saving technology, they are able to preserve water. Thus, it was a question of financing, not of availability, whether the farmers agreed to support the nomads with water or not. The nomads could also make an interesting offer. Manure is indispensable for irrigated agriculture and the farms have a demand for manure that could not be satisfied by local production. Hence, the nomads started to supply the large farms with manure for good prices. Other subsidized commodities were also exchanged for water and implicit support, which were of particular interest for the wealthy farmers, among them fuel (mazout). Thus, a new complex system of exchange was introduced to the detriment of the local peasants. Included in the contract is that the nomads never cause problems for the large property holders. This was the hour of the construction of a new strategic group composed of members of the farmer elite on the one hand and the owners of large camel herds, respectively their local representatives, mostly Sahraoui clan chiefs, on the other hand.

By comparison, very limited benefits from nomadic production remain today for the locals. The nomads no longer contribute to stable prices on the meat market. On the contrary, based on their possibility of moving the flock, they have much better chances to profit from price fluctuations than others. Moreover, most of their herds disappear in the south which means in concrete terms that the meat marketing options are associated with contraband trade networks of neighboring countries. In short, the profits of radical overexploitation strategies in the Souss are siphoned off somewhere else. However, some vestiges of ancient complementary relations persist. Local religious festivals (moussems) are the occasions for which the Sahraoui still today provide the necessary camel meat. But in general benefits are now reaped by others.

In the course of time the nomads also began to take up local occupations, which are actually associated with the peasants' way of life, increasing competition. For instance, they started to produce Argan oil. But most particularly they started to produce manure and to compete with local resident suppliers. Additionally, the long distance mobility of the herds is no longer indispensable. In fact, mobility in general became a strategic factor in a complex economy. The favorable ecological and social conditions motivated nomads for longstanding residence as mobile pastoralists in the Souss plain at least for most parts of the herds and members of the group. They only occasionally visit their former homesteads in the south.[16] They are installed even in areas, which in the past were not suitable for longer stopovers because of the lack of water. Today, however, it is not the herd which comes to the water; it is the water which comes to the herds by way of cistern trucks, allowing the overexploitation of even the remotest parts of the forest. But also the animals are carried with trucks for longer distances. The economy of cyclic mobility backed up by arrangements with local residents has been replaced by a modus of ad hoc decisions about mobility, which includes the exploration of eventual pasture grounds with new information technologies. Since the mentioned contributing factors – water saving technologies, cash-crop production in the forest etc. – have radically changed the situation since the late 1980s, fixed arrangements with particular local sedentary partners are no longer essential. The network with farmers is much more flexible than traditional arrangements ever could have been. Furthermore, the new strategic occupation allows the maintenance of herds of a size which was hardly possible before.

Summarizing the economic factors, the competition refers to the access to natural resources, namely pasture in the forests and water. A further economic criterion is the supply with subsidized commodities such as fuel (mazout), food, and other tradable goods. Wage labor, manure production, the meat market and the usurpation of local occupations, such as Argan oil production, are contributing factors.

Legal Arena

The legal praxis in the juridical field discussed here is specifically referring to a variety of legal reference frames. There is a lack of state regulation on the one hand, particularly concerning the integration of the Sahraoui economy in local environments. Open access and freedom of movement were granted by the Moroccan state for the Sahraoui after the *Marche Verte*. Access to pasture for the nomads has never been clarified in a state legal order. There is no legal backing for the whole spectrum of practices of resource exploitation in the forest since the forests have been completely put under state control during the French protectorate. Sharecropping, for instance, is performed as a moral economy based on a local folk model. In the past, contractual relations between local residents and mobile pastoralists were concluded on the basis of exchange and complementarity. Some of the established rules persist. When asked about the acceptance of local rules of coexistence, the herder's answer was mostly affirmative: "Yes, of course, even today we are still obedient to our verbal agreements. Have you ever seen, for instance, one of our camels straying in the villages?"

As it has been said, a vital question was how to regulate the access to water. Aquifers were reserved for irrigation and nomads had no legal access to wells. Wells remain individual private property even today. Thus, access to water was dependent on agreements and the new arrangements of payment for water affected both groups of actors at the same time. The connection of the region with the world market transformed water into a calculable investment factor which no longer seems to be dependent on God's will. Thus, the sophisticated traditional legal orders regulating access to water for individuals and collectives gave way to increasing privatization of access to natural resources.

Other regulations prescribe the moment for opening the pasture season after their closure during the cultivation period. This was traditionally after the harvest of cereals and the collection of the Argan fruits. But then, with the extension of irrigated agriculture into the forest, there was no longer a concrete period that could be defined as 'after harvest season'.

The state adopted elements of the customary law (ᶜurf) in legislation and in this way made it inflexible. It was officially confirmed that the population of the neighboring villages has rights to use individual plots in the forest.[17] The plots are inheritable and, accordingly, split upon inheritance. Official inheritance rules, whether state or Islamic prescriptions, have only a limited value. Within communities of heirs, access to an inheritance was quite often monopolized by a prominent and respected elder male community member. Different types of landed property, including collective or communal land, may be forested, while the forest as such is under state control and individuals or communities are only entitled with usufruct rights. But officially, except in the closed season, pastures are openly accessible. The communal use of the forest as pasture has been held responsible for overexploitation.[18] But communal use is not always practiced. Furthermore, in practice, the combination of exclusive usufruct rights with communal access to pasture has led to a control of the generic use of individually ascribed plots.

To sum it up, confusion reigns. State intervention and state control contribute to the legal insecurity for local users: "There is no law to protect us. That's why we decided to reactivate our *ᶜurf.*" Competing and contradictory legal regulations regulate practice. Within this setting, the exertion of low level violence in order to show the ability to claim and maintain access rights is an accepted option in the local context. Interpreting these traditional rules strictly, then, there is only very limited space for the escalation of violence. In a particular case of conflict between a resident peasant and a group of nomads, who set up their encampment against the peasant's will on the plot where he is exercising his rights of usage, his neighbors told me: "Well, he has God's blessings and will not concede. He is father of five adult and strong sons and one of them is even a butcher".

Intervention by development agencies and other external actors such as the UNESCO contribute to the complexity of the situation. Transnational juridical standards of resource management according to criteria of sustainable development and ideas of universal environmental protection on the basis of international conventions should, in theory, be adopted into national law.[19] The fact that this process did not produce the desired results motivated actors in development cooperation and powerful donor organizations to promote the implementation of these standards immediately in the local field. This was undertaken partly with reference to local legal ideas. Revitalization of tradition was seen as a transmitter of an ideology of sustainability and a western perception of the environment. New rules have been propagated as the revitalization of 'good' tradition, such as *agdal*. This has been understood as a remedy against the nomads: "They do not let us do our job undisturbed. That's why we decided to return to the *agdal*. It's a good idea and the *barrani* (foreigners) promised to support us," as an adherent of a local environmental NGO, supported by a development agency, declared.

But development agencies do not promote a homogenous approach. Transforming the forest into a resource in the fight against rural poverty and at the same time preserving it as a unique ecosystem hardly seem to be compatible goals in practice. With allusion to local rules, transnational actors promoted new inventions. Fencing in the forest with rows of cacti as a measure against desertification (PAN) is just one example. This is incompatible with the basic principle of open access to the forest. The strategy of fencing is also a good example for the state of legal insecurity local users are facing. Some forest areas have been selected for reforestation projects and are therefore fenced-in. Those who have use rights there lost them without any compensation. Simultaneously, the pressure on the accessible parts of the forest is increasing.

Under these circumstances, the local informal institutions of conflict settlement have been pushed into the background. The owners of small herds still showed their principle willingness to negotiate and to compromise with the locals by means of intermediary institutions. The larger herdsmen, however, just said that they had no permission from the absent owners to compromise.

Mobile Pastoralism and the Ideology of Being a Nomad

When looking at the transformation of nomadism in its relationship to resident groups or the development of mobile pastoralism in general, the risk is that other dimensions of nomadism are neglected. Here, I will focus on additional aspects of nomadism, on 'the ideology of nomadic mobility'. In the following, I will refer exclusively to the empirical data about the Sahraoui, which cannot be generalized for all mobile pastoralists who once were active in the Souss.[20]

The economy of nomadic pastoralism is more than herding. Mobility is one of the cornerstones of the nomadic way of life and the ideology of nomadism is more than a reflection of economic activity.[21] Even when sedentarization occurs, the ideal of mobility is still maintained as a metaphor of freedom and independence. On the social level, the ideology of independence through mobility is connected with a strong sense of social cohesion and the maintenance of kin solidarity over long distances. Mobility such as labor migration, for example, may be associated with the herders' ideology and regarded from a nomadic perspective. The nomadic ideology also includes professional skills of the exercise of violence. Protection of trading routes, if not armed raids have been practices in the past that are still remembered vividly today. The nomadic warrior ideal made nomads perfect candidates for military service. Trading is also included in the spectrum of nomadic activities. In the past, long distance caravan trading was practiced and the nomads provided the means for transportation and participated in exchange. The combination of military skills and trade remains one of the most important aspects of the nomadic economy of the Sahraoui in the Souss today. Subsidized commodities from the south find their way back to other parts of the country, in the concrete case, particularly *mazout*, agricultural materials and food. Today, trading partners of high political and social rank such as holders of large-scale landed property in the described context are included in the commercial network.

In the concrete case of the Sahraoui camel nomads, they have only one highly specialized and esteemed branch of activities, namely, camel breeding in combination with huge goat herds. Herding goats and sheep may even be far from prominent in the rank of nomadic virtues, but the highest ideal value does not necessarily designate the most profitable field of occupation. Moreover, camel herding is embedded in a bundle of other activities. This gave rise to the social construction of a quite modern type of nomad who still refers to that ideology correlating a range of economic activities with social values as well as genealogical and social solidarity. In the context of a neoliberal economy, nomadic virtues still constitute a type of social capital and a resource, which can be profitably exploited. This fact is obviously considered by the external and non-nomadic investors in camel herding.

Therefore, the ideology of nomadism, as the foundation of the identity and self-perception of the mobile groups concerned, provides an analytical basis for identifying markers of social cohesion and distinction, which allow for maintaining relationships to allies and determine the lines of confrontation beyond the criterion of mobile herding in a system of exchange. There are markers of distinction emphasizing nomadic identity: living in a tent, for instance, or being a bad Muslim. Nomads do not respect religious food prescriptions, drink the blood of birds and camels, and eat animals such as rats, turtles and lizards and they do it proudly.

It is not self-evident, how these aspects of nomadism may be connected with discourses on the national level in Morocco. Two parallel and disconnected lines of narratives dominate the public discourse on nomadism in Morocco these days. There is, on the one hand, the reemergence of romantization and exoticization of nomadic lifestyle in the public debate as propagated in the media. It would be simplistic to reduce this tendency to a strategy for the attraction of tourists. It obviously is tied to the reorientation towards local traditions and the revitalization of tradition as an intellectual resource and property connected with the construction of a pan-Moroccan identity. There is more behind the reestablishment of *nomadic moussems*, for instance, than just creating more opportunities for entertainment.[22]

On the other hand, nomads have a particularly bad reputation due to the fact that they are traditionally held responsible for environmental degradation in Morocco (Davis, this vol.). Furthermore, the association with violence and repression and cultural and social inferiority expressed in numerous jokes about the illiterate and incredibly putrid-smelling Sahraoui target the mentioned markers of distinction. Other narratives emphasizing nomadic virtues as simply uncivilized behavior target gender relations, rape, kidnapping and similar issues. Moreover, the political signals from the central government are perceived as inconsistent whether the Sahraoui are privileged and even immune or integrated in Moroccan society on an equal footing.[23]

Conflictual Relationships

Time and again conflicts have arisen over access to the forests, but since 1996 they have become more and more serious. The invasion of huge camel and goat herds from the south blocked access to pasture for the villagers who actually had collective use rights to them. In fact, pasture is free of charge, but this rule cannot be interpreted as an invitation to completely destroy the resource. Furthermore, the Sahraoui no longer feel bound to respect the basic agreements of coexistence. They not only invaded the stubble fields in rain-fed agricultural land (bour), whose exploitation is the right of the locals, but also destroyed cereal fields before the harvest could be brought in. Further, in several cases they refused to negotiate their date of arrival, despite the fact that the harvest and the Argan fruits had not yet been collected. In one case, the herders who unexpectedly arrived with a large camel herd threatened: "If you keep us away from pasture, thumping outmoded restrictions, then we'll explain to you with the slingshot that our masters are the most powerful". Then, the situation became explosive. In several cases camel troops entered sharecropping fields where expensive cash crops were cultivated and devastated them. These fields are fenced with windbreaks so that herdsmen can easily keep off animals. The losses for the individual sharecropping families were considerable. Their field neighbors, however, acknowledged their responsibility as mutual assurance group and contributed to a partial compensation for the victims.

A permanent state of tension between both parties developed and politics of permanent challenge and response between the groups emerged. The peasants who complained about their losses before the Sahraoui chiefs were severely beaten up and threatened that they would experience even more problems, if they continued to be annoying. Some sharecroppers have been attacked by herdsmen on their own land while working their fields. Women and girls refused to assist as seasonal harvest laborers, horrified by the Sahraouis' reputation as 'professional rapists'. Furthermore, the Sahraoui signaled that they were being backed by high-ranking state representatives, conferring them a sort of legal immunity. In reaction, the informal village council of the concerned community decided to authorize some of the youth to kill and continuously steal animals from the nomads, as compensation for the devastated fields and in order to terrorize them. In some cases they even poisoned the nomads' water reserves. In response, Sahraoui started to cut down trees, partly for pasture and firewood but also in order to provoke the actual executors of the exclusive usufruct rights.

In 1998, youths from a neighboring village attacked one of the encampments and demolished the tents. The nomads were forced to withdraw but not without announcing their return. In 2001 the situation escalated and several persons were severely wounded, and one was even killed. The *moqla* (slingshot) became known as the weapon of the nomads. This event went down in history as the 'battle in the Argan forest of Ait Tigemmi'[24]. A Moroccan development broker described his impressions of the event as follows:

> I couldn't believe it. I felt like being on another planet. It was a medieval and at the same time hypermodern type of battle. Their [the Sahraouis'] women stood up on the hill crying youyou and 'dancing' [jumping on the spot with naked breasts]. Meanwhile, the men, armed with clubs, slingshots, knives but also jeeps that they used like tanks in order to disperse their enemies were organizing their action using mobile phones.

The indecisive attitude of state representatives in monitoring and moderating the tensions did not contribute to its pacification. The persistent complaints of the peasants remained without redress. It went even worse for them. While the persons responsible for the cutting of trees have never been arrested, either by the forest rangers or the police, some of the local rustlers have been. The villagers understood this as open partisanship on the part of the state actors. The arrested villagers have been released at the expense of the whole community. This was, by the way, the occasion for a big release party. All in all, most of the state functionaries, the *qaid*, the *sheikh*, the *muqaddim*, notables who were involved felt uncomfortable with the situation and insinuated that their hands were tied by a higher order. The state's credibility decreased accordingly. The lack of state support for local residents found expression in a mixture of resignation towards the state and hyper-nervous reaction towards the nomads. All along, the big farmers, who do not regard the survival of the sharecroppers and small peasants with enthusiasm, continued to deliver water to the nomads and to do business with them.

When traveling in the Souss today one quite often encounters complaints about the increasing tensions and the rising propensity for violence. Violent exchanges even take place in the form of arranged battles under state observance, especially during the summer season. Conflicts and violent events are openly discussed and occupy a prominent position in local discourse. State representatives, local NGO's and other representatives of civil society comment on these events regularly.[25] The meta-dimension of the conflict potential and the social lines of conflict behind them, however, are not addressed publicly and are not even perceived as being part of Moroccan social reality. Everything is put in the context of a nomad-peasant dichotomy.

Transnationalization of Local Economy and Social Dynamics

The outlined coincidence of agricultural and political developments together with the transnationalization of the local economy resulted in a social stratification and unequal distribution of wealth and resource access rights. The emergence of a post-protectorate political elite imposed new social relations and a new order and delegated the resulting social drama to the disempowered former partners in resource exploitation, to the detriment of existing complementary relationships.

New farmer-pastoralist networks continued complementary resource exploitation and established a system of exchange, which went far beyond the so-called conventional *rahhalin-fellahin* cooperation, inventing new dimensions of mutual support. This model, which is informed by transnational impacts, relies on the renegotiation of the allocation of political and economic power and – from a nomadic perspective – proved at the same time its compatibility with the nomadic ideology. The networks were successful in combining horizontal alliances on a high level of social stratification with vertical ties up to the highest class of political authorities. The established new cooperating network relations are socially expressed in mutual invitations on various occasions, such as *moussems* or local festivals. I accompanied farmers from the Souss to Guelmim on one of these occasions, which ended in a sort of political demonstration.

In reaction, new networks made up of small peasants who were forced into the role of the conflict partner of the pastoralists, while the latter became cooperative partners of the landholders, emerged recently. After 1996, external intervention incited the installation of local supporter groups. New actors appeared on the scene and contributed to the delineation of conflicting parties, both of which proved to be highly effective in alliance building.

Social dynamics have been negotiated in the nomadic discourse, which has been associated with the sensitive political issue of the West Sahara conflict. Social tensions and contested power relations have been negotiated in terms of nomadism and mobile herding. The locals, however, roundly refuse this kind of argument in declarations which demand equal treatment: "This is one country for all of us and we all belong to one and the same king."[26]

Social Lines of Conflict

At first glance, the lines of conflict and violence between groups of actors, which can be identified as local *fellahin* on the one hand and mobile *rahhalin* on the other, seem to confirm the theoretical model of a shift from complementary exploitation of natural resources and exchange of goods to conflict. But behind this conflict, on the level of inter-elite cooperation, there are neo-complementary groups of actors in the Souss, described here as strategic groups.[27] This implies not only alliances for mutual profit from economic cooperation, but also from the new conflict potential their cooperation imposes on others. At the same time the partners downplay the social significance of these tensions generated through their elite cooperation.

Furthermore, the new constellation cannot be approached as a shift from one prevailing model of contractual relation between two social partners to another model with two other cooperating partners to the detriment of the former partners. The analysis shows that the social process cannot be explained as a transformation of a double dichotomy. It is an asymmetrical constellation that has evolved. The dependants of the prosperous owners of large camel herds, who are mostly destitute, hired herdsmen, are opposed to the pauperized opponents of the wealthy landholders. The latter are cooperating with the owners of the camel herds, who at the same time hold positions as clan chiefs disposing of political power. Most of the herdsmen I asked were either affiliated and dependent members of their masters' kin groups or in a sort of clientele relationship. The masters, however, actually act as intermediaries. Some of the real herd-owners in the background take great pains to remain anonymous. In comparison within the relationship between the herdsmen-herd and owner relationship, the peasants have relatively more room for maneuver vis-à-vis the landed property owners and at the same time compete with them over resources.

In short, the exercise of violence is delegated to particular social groups who need to fight for survival while the other actors are exempt from exercising violence. So, in terms of strategic grouping, one recognizes the alliance of three groups of actors against one remaining group. This is what is meant by an asymmetrical constellation. The herd owners, the landed property owners and the poor hired herdsmen as executors are allied against the local peasants. On the side of the losers of these developments there also are the former seasonal partners of the peasants practicing diverse forms of transhumant mobility. The longstanding relations between these partners have been pushed aside although they continue to a certain extent, particularly in regions that are less attractive for the camel nomads.[28]

The constellation corroborates the hypothesis that complementary relations for the achievement of common objectives demand a state of equity between, in the concrete case, the two groups of elites. This social type of equity is based on the state of legal insecurity restraining the maneuverability of the other actors involved.

In this constellation, both sides were looking for allies and found them. The Sahraoui have included in their network an intermediary group, whose members are able to communicate with both parties, the local Sahraoui nomads living in the Souss since their arrival from the south in the 1970s. These people had sold their camels and invested in agriculture. Realizing that this, still, will never be an acceptable alternative way of life for them, they began to rent their land to local peasants and limited their own activities to the keeping of small livestock. In addition, they were successful in accessing a new form of labor, namely, construction work, where they act as mobile teams of workers. The satellite settlements and camps founded in the 1950s, 1960s and 1970s still serve today as refuge areas for their nomadic relatives in cases of conflict escalation. Furthermore, when expecting a violent event, the Sahraoui may hire supporters on the construction labor market under Sahraoui control.

After 1996, the peasants on the other hand, for the first time also had a powerful partner available to them, namely the transnational actors. These interveners regarded the local peasants as the target group for the establishment of transnational standards of sustainable resource management. As mentioned, demonstrating the ability to exercise low level violence is generally accepted as the appropriate way to maintain access rights in a local context. Now, escalation has taken place because both sides feel they are being supported by strong allies. Insofar, transnational actors did not introduce new dimensions of conflict but contributed to the reactivation of local interest in the violent option (Turner 2005).

Conclusion

The case study reveals how fragile systems of exchange depend on the variety of crops produced, land rights, ecological factors as well as state and transnational intervention and the degree of integration of local economies in the global market.

The complex relationship between the different but interrelated networks provides an analytical tool for the understanding of the new dimensions of violent conflict in the Souss. This aspect has been overshadowed in the conventional way by implicitly emphasizing the shift from complementarity to confrontation as the typical result of increasing competition in resource exploitation.[29] Furthermore, the described displacement competition sheds light on the processes of disengagement of local elites from their local social environment. They do not feel a binding responsibility to the local community as a whole for which they used to keep political functions as elected representatives. The professional connections the farmers maintained with the nomads, and particularly the provision of water, changed the peasants' minds. The reaction of the peasants can be described as a formation of local resistance against the elite's hegemonic claim to accessing increasingly scarce resources. This phenomenon may be interpreted as a reaction in conformity with the model of strategic grouping in resource competition (Berner 1995). The stereotype of the eternal nomad-peasant conflict disguises accumulative processes concerning political and economic power in the rural countryside.

Looking at existing development programs, it cannot simply be a question of integrating the mobile pastoralists or not (Werner 2006). Targeting rural poverty and rural development implies integrating all social actors in one concept of action. Accordingly, neglecting the real alliance of nomadic and peasant elites as a special local arrangement will bring about a collapse of development programs.

Notes

(1) See Diallo and Schlee (2000, especially van Dijk). For the Souss see Aziki and Bodemeyer (2003, 18).

(2) A strategic group is defined in terms of shared interests in the extension of economic success, but does not necessarily include concrete contexts of concerted action or the development of a cohesive collective identity. For further details see e.g. Berner (1995).

(3) See e.g. Benhaddou (1997) on elites and power relations.

(4) According to the model Hirschman (1970) developed of organizational decline, members of any kind of social formation may operate with two essential options, i.e. 'exit' and 'voice'. In the concrete constellation, those actors pushed to conflict could not

withdraw (exit) since they depended economically on allied actors of high rank. Instead, they had to realize their voice option against the direct opponent violently. For a pastoralist-agriculturalists context see Dafinger and Pelican (2002) referring to Hirschmann (1970).

(5) Cf. Swearingen (1987).

(6) Many concomitant aspects, which contribute to the dynamics addressed in the paper – such as a tremendous demographic growth, economic factors of migration, tourism and particularly religious factors as the intervention competencies of local religious experts in conflict matters – will not explicitly be discussed here.

(7) Montagne (1930 [1989]); Hatt (1996).

(8) Cf. Montagne (1930 [1989]); Jacques-Meunié (1982).

(9) All literal quotations in this paper are translated from recorded interviews and discussions with local informants – villagers, peasants, farmers, nomads, development brokers and state officials – held during fieldwork in the years 1996–2005. Empirical data presented here refer to a rural community located in the center of this region where a particular combination of factors is given, singling out this site from the high variability of natural and social conditions. The data on which this paper is based have been anonymized. Fieldwork on transnational impacts on local social dynamics between resident rural population and mobile pastoralists was carried out for several weeks annually between 1996 and 2005. Data from different settings and concrete constellations in the Souss plain are included in the analysis. Since 2001 the fieldwork has been part of a project of the Legal Pluralism Project Group at the Max Planck Institute for Social Anthropology in Halle, Germany, on Sustainable Development and Exploitation of Natural Resources, Legal Pluralism and Transnational Law in the 'Arganeraie' Biosphere Reserve.

(10) For examples see Werner (2006).

(11) Aziki and Bodemeyer (2003, 11–14).

(12) For details see Popp (1983); Swearingen (1987).

(13) From conversations with sharecroppers, 2000.

(14) This is only one of many reasons for the degradation of the forest and particularly true for the concrete case considered here.

(15) It would be worthwhile in this context to analyze the political dimensions of 'drought' as a concept as well as its defining character and its argumentative power in the interaction between sedentary peasants, nomads and state representatives in connection with the discourse on high risk economy linked with the world market and high rates of profit. See Werner (2006) for some functional approaches to drought in connection with nomadism.

(16) Further literature on pastoralism in the Souss in Werner (2006).

(17) For an overview of the particular forest legislation see RBA (1997/2001).

(18) See e.g. Aziki and Bodemeyer (2003, 7); Werner (2006). One-dimensional and simplistic explanations are, however, contested. See Davis (2006) for literature. There is no place here to go into details on the 'tragedy of the commons' issue under neoliberal and transnational conditions. The plots in the forest are neither common property nor openly accessible. The described case reveals one of the typical constellations where individual and exclusive usufruct rights to plots combine with communal claims to specified access. Processes of incremental appropriation (melkisation) in favor of local holders of usufruct rights collide with interests of Sahraoui nomads and large-scale landed property holders as well.

(19) The principal legal set of relevant regulations is laid down in the UNESCO program 'Man and the Biosphere' and the UN program to combat desertification (PAN/LCD). The wider framework has been set up in international conventions such as the 'Rio Convention' of 1992 (CNUED-Rio 1992) and the 'Convention to Combat Desertification' (CCD-Paris 1994). Cf. RBA (1997/2001); MADR (2001); UNESCO: Réserves de Biosphère: La stratégie de Séville & le cadre statuaire du réseau mondial, 1995. http://www.unesco.org.mab. See also http://www.un.org/esa/agenda21/natlinfo/coutr/morocco/natur.htm; http://www.unccd.int/. Since 2001 PAN/ LCD has been realized as a *programme d'action nationale* in Morocco (MADR 2001). For the wider context cf. Davis (this vol.).

(20) There is no space to comment on the general discourse on the nomadic way of life and what constitutes a nomadic society; cf. e.g. Salzman (2004).

(21) On nomadism as a concept, habitus, and ideology see esp. Vajda (1968).

(22) One of many examples: Nouiga (2003).

(23) This has to do with the sensitive issue of the Western Sahara, integrated into the Moroccan kingdom as Southern Provinces while still on the United Nations' list of 'Non-Self-Governing Territories'. For the ambivalent social position the Sahraoui enjoy between promotion and restrictive state control see e.g. Le Journal Hebdo N°206, (2005).

(24) This name has been invented by the author.

(25) Cf. e.g. the webpages http://www.tanmia.ma/article.php3?id_article=1539&lang=fr or http://www.arabmab.net/biosplan.cfm?bid=15; accessed 04/11/05.

(26) This is – with variations – a common saying in the whole of Morocco.

(27) Evers and Schiel (1988); Berner (1995).

(28) See e.g. Werner (2006).

(29) See e.g. http://www.tanmia.ma/article.php3?id_article=1539&lang=fr.

–14–

Nomads: But how?

Hassan Rachik

The Beni Guil

Spatial mobility has always been the strategy of nomadic groups favoring their independence in relation to the state and to sedentary groups. In the case of political or climatic instability, the tents could be packed up immediately and the herds led into a more peaceful and clement environment. The perpetual mobility in search of pastures with better vegetative cover remains the major advantage of nomads, their reason for being. Considering the aridity of the high plateaus of the eastern Moroccan steppes (210 mm/year) and the immensity of the confederation's territory (more than 26,000 km^2), the Beni Guil are frequently forced to fold up their tents in search of new pastures.

The collective name Guil refers neither to a common ancestor, nor to a common territory. It rather resonates legends, which reveal essential traits of the nomadic life, for example, the choice of the tribal territory. In the legends pastoral activity is highly esteemed and often valued higher than tribal adherence: it's the grass that counts. In comparison to a green pasture, the sentiment of mere tribal allegiance means little.

For the Beni Guil spatial mobility was never an option, but a necessity. Without it, any exterior relationship would have been impossible, whether the conquest of new spaces, economic exchanges, or caravan commerce, to name but a few. Until the 1930s, no market was organized in their vast territory. For the sale of their pastoral products and the purchase of basic goods like clothing, cereals, tea, sugar, they had to move towards the markets of neighboring or faraway cities: Figuig, Beni Ounif, Debdou, Melilla, among others.

Since the 1960s, spatial mobility is nothing more than an alternative.[1] The tent, which still constitutes the habitation for the majority of the Beni Guil (65%), stopped being a sign of mobility. Many pastoralists bitterly remark that they are no longer nomads, *mabqinach rahhala*, even though they continue to live in tents, far from the urban centers. The local fixing of the tents, their immobilization, is however a recent phenomenon. Of the 49% of nomads affected, the majority (78%) avowed to have abandoned spatial movements after the droughts of the 1970s. The pastoralists, who continue to move and who perpetuate nomadism, adopt innovations but had thus to pay by breaking with old practices. The most manifest and impressing rupture is the abandonment of the dromedary, the traditional symbol of nomadic mobility. Since the very beginning of my research, I felt that this rupture is the starting point for the description and analysis of the transformation of nomadism and of spatial mobility.

Whenever the term 'nomad/nomadic' is used to qualify the Beni Guil, it is not to define the group in its entirety, venturing from its relationship to space, but rather it is aimed at emphasizing one fundamental morphological dimension of their social organization. Nomadism cannot characterize an entire society, since it is affiliated with several forms of political and social organization. This paper will show that among the groups studied, mobility was organized in the framework of distinct, even opposite social structures, some based on contracts implying egalitarian values and norms, others founded on dependence and patronage, denoting values tied to charity. Nonetheless, whatever the social and cultural framework may be, mobility has similar effects on the political and cultural life of the nomadic groups.

In analyzing conversations with nomads, two terms implying the notion of nomadism stood out: *rehel* and *thawwel.* The verb *rhel* locally applies to movements of great amplitude, 30 kilometers or more, undertaken by a pastoralist, and the verb *thawel* to a change of campsites in a 2.5 to 5 kilometers radius within the same pasture ground. The term *rahhal* thus designates somebody, who moves and changes the pastures during the year. In contrast, the word *mgaytan,* from the word *gten* is employed to signify a nomad's immobilization. He is the one who only moves his tent for a change of campsite. Being *rahhal,* means having the habit of moving with one's family in the search of water and pasture. It is in this sense in which the word "nomadity" (rahla) is used here. It designates the mobility of a family or of a group of families during the year for reasons connected to pastoral activities. In contrast, the term "transhumance" refers to a seasonal mobility that only implies the movement of some family members, sometimes even only of shepherds.

Compared to the notion of *ᶜrab,*[2] which characterizes rather a culture than a mode of relations to the soil, the notion of *rahhal,* referring to spatial mobility, is less complex. The difference between these two notions is however essential. A nomad, who has permanently fixed his tent, asserts that he is no longer *rahhal,* thus underlining a change in the spatial dimension of his identity, but he still affirms his adherence to *ᶜrab,* accentuating his cultural affinities. A Bedouin, who fixes his tent or installs himself in town, continues to share values, perceptions of social relations, knowledge of livestock and environment with his nomadic brothers. Being *aᶜrab,* just like being *fellah* or peasant, is a way of life, a way of connecting to the world and to society, a culture. This distinction is paramount, because the changes that I will examine here, with regard to the extremely limited time that was given for the study on the terrain, principally concern the Beni Guil as nomads. I will address norms and values and their relationships with the daily organization of pastoral production. In other words, cultural structures are not the focal point in this paper, but rather social processes, their founding norms and values. For example, the disappearance of certain values, like charity or mutual support, and certain norms, like pastoral contracts and the adoption of others are examined here in the framework of the organization of mobility and pastoral production.

Moving the tent and the herds was more a collective task than a family affair. Traditionally the networks of social relations have been organized due to the demands of spatial mobility and pastoral production. The respective pastoral communities rarely exceeded ten households and were not 'formally' defined in referring to genealogies or normative divisions of the confederation in tribes, fractions and lineages, but originated rather from pragmatic considerations. Here they are understood as 'concrete groups' in the sense that they used to constitute a social network and an economic association in the frame of which herd and human mobility was effectively organized. They were flexible and fluid, because they formed and dissolved for primarily pastoral reasons. It is at the level of these restrained communities that the changes, which have affected spatial mobility, are observable, and at this same level that the innovations of nomads and their relationships to structural and contextual changes – concerning adaptation, commercialization, conversion, and survival – will be discussed and analyzed.

Adaptation

The recent transformations – name it globalization – cannot be reduced to a mere series of constraints which disadvantage nomadism. For they are also a set of resources and opportunities that permit the nomads to adapt to changing conditions. Adaptation here denotes the act of benefiting from these new resources in order to perpetuate nomadism in innovative and unprecedented forms.

Whatever the daily efforts expended in the maintenance of livestock may have been in former times, it seems as if a nomad did not conceive a direct link between pastoral production and human effort. In his mind 'the herds grew naturally'. Among the Beni Guil, only the horse, a prestigious animal held by the wealthy, is nourished by barley. The fodder of all other animals was provided exclusively by pastures and no supplementary fodder was provided. Our interlocutors regret the vanishing of the old times when the herds grew naturally. If there exists such a thing as a painful innovation, then for them this undoubtedly is the introduction of external forage. Speaking of changes that have affected their society, not one of them omitted to mention the introduction of forage. "Formerly, stock grazed with its mouth, today we spend money to feed it," as one of our interlocutors put it (Rachik 2000).

The introduction of forage crops has become a focal point of reference in local history. Two dates are retained: *ᶜam chmandar*, the "year of dried pulp" in 1978/79, and the *ᶜam coudissa*, "the year of coudissa", named after a kind of industrially manufactured fodder introduced in 1989. The retreat to external forage rapidly spread after the droughts in the 1970s. The nomads started by feeding part of their herds, notably pregnant or hurt ewes. From the 1980s onwards, the habit of forage supply massively gained ground, in other words entire herds were given supplementary fodder throughout the year. In 1991 the majority of nomads, that is 80%, practiced some form of supplementation.

The nomads however did not always appreciate the use of forage crops, which was seen as a shame (ᶜayb). It was considered ridiculous to provide the herd with supplementary fodder, instead of simply leading it to the pastures, which at that time were abundant. Supplementation was rather viewed as burlesque behavior, as it challenged the very foundation of nomadism: the perpetual search of 'the grass of God'. Nomads, who did not rely on the pastures and who dared to interfere with the natural growth of herds, became isolated. However, with the shrinking nomadic territory and after the degradation of pastures, aggravated by the drought of 1970, what used to be a shameful practice became a dire need.

It is quite common to explain pastoral innovations as related to droughts. Yet it must be asked why, in a region that is generally prone to droughts, it is precisely that of 1970, which triggered the massive introduction and adoption of external fodder. It has been illustrated that nomads who can no longer successfully apply traditional coping strategies are those who are willing to use new pastoral resources (Rachik 2000). The supply with forage remains a solution of substitution, but also an opportunity enabled by the new political landscape of the independent Morocco. Since, it is the state's duty to assist the nomads, particularly the Ministry of Agriculture's duty, for example, to subsidize the barley price.

Proximity to administrative services has become a principal resource to sustain nomadism and consequently a local concern. The administrative partition of 1976 bisected the Beni Guil in two rural communes, each administered by a *caid*, who represents the Ministry of the Interior on the local level. Gone is the time, when the *caid* was a nomad himself and, following his herds, had decision making power over the steppes. To emphasize the immobility of the new functionary and to avoid all ambiguity, he is called *caid lbirou,* chief of the bureau. It is the function of a sheikh, representing the *caid* locally, which has become a capital game. His unit of command, the *Mchikhat*, offers the ultimate administrative frame for the distribution of subsidized fodder and for the concession of other public services. Until the creation of cooperatives in 1991, it was in collaboration with the sheikh that the technicians of the administration determined the quantity and the quotas of food items allocated to nomads.

In the wake of Morocco's independence, all tribes of the Beni Guil were grouped into six *Mchikhat*, according to the profile of the ancient *khoms*. At their heads, the Ministry of the Interior assigned a sheikh, who adhered to the nomads. He was not a functionary, but received indemnities of 700 Dh per month from the Ministry.

Table 14-1: Innovations and Their Adoption Among the Beni Guil

Innovation	Period of adoption (% of nomads interviewed)						
	before 1970	1970 - 1975	1976 - 1980	1981 - 1985	1986 - 1991	without date	Total
Abandonment of displacements	19.7	27.3	14.2	19.6	16.7	*2.2	100
Selective treatment of ewes	16.8	35.2	11.7	11.0	25.4	--	100
Abandonment of horses	14.0	33.8	12.2	10.7	29.2	--	100
Troughs	13.4	33.7	22.6	17.6	12.8	0.7	100
Transported water	8.6	37.9	18.2	16.6	13.1	6.6	100
Trucks	6.2	12.4	19.5	27.9	34.1	--	100
Fences	6.1	34.2	20.2	16.2	20.6	2.8	100
Pulp	2.2	22.8	40.1	20.5	12.6	1.7	100
Antiparasite treatment	2.2	9.6	16.3	21.8	47.4	2.7	100

Source: Rachik (2000). Remarks: * = unknown

In order to attain access to new resources drained by the reformed administration, several tribes demanded their own *Mchikhats*. This was particularly the case for less prominent or subaltern groups, which did not furnish a sheikh. As a consequence new *Mchikhats* were created, which broke up with the old tribal divisions. In the aftermath three new autonomous *Mchikhats* were founded before 1970, seven between 1970 and 1982, and 16 after 1982. It is at the *Mchikhat* level that pastoral cooperatives have been created in the early 1990s by a development project for pasture and pastoral amelioration under the auspices of the Provincial Administration of Agriculture (Direction Provincale de l'Agriculture, DPA).

A convention, which I attended, illustrates at which point the relationship to the administration has become important for the nomads. It took place on April 18th 1991, at the office of the rural commune of Tendrara on the initiative of the livestock service of the DPA, whose office is situated at Bouarfa, 60 km away (cf. Mahdi, this vol., Map 6-1). All local decision makers were present, the chief of the livestock service, the representative of the *caid*, the presidents of the pastoral cooperatives, the sheikhs and notables of the concerned tribes. An item on the agenda was the definition of criteria regarding the distribution of subsidized fodder. In the course of a heated debate, several regulations were proposed: These included the reintroduction of a former regulation, which attributed equal parts to seven of the *Mchikhats* and smaller shares to two, the Oulad Mbarek and Oulad Ali Bellahcen, who hold less animals. Competing proposals called for the determination of a quota according to the total number of sheep held by each cooperative, or according to the value of shares possessed. *Mchikhats* with low animal stock defended the reintroduction of the first regulation, while those owning large numbers of livestock demanded a rule of proportionality. The sheikh of the tribe Oulad Hmad was the first to propose and support this latter rule. His colleague and accomplice from the Oulad Fars reminded that his tribe, since the colonial period, had always been considered as a grand tribe and thus believed it unjust that fodder should be distributed without taking the groups' respective importance into account. The assembly ended without stipulating a consensus between large and small tribes.

The nomad is thus not any longer alone in time of crises. The dilapidation of traditional strategies is buffered, for example, by the possibility of acquiring external fodder even in the middle of the steppes. It also has become impossible to rely on tribal structures alone, hence the nomad has learned to lean on the administration's services, which are obliged to assist him. The DPA in return organizes campaigns of vaccination, anti-parasite baths, and – in agreements with the nomads' representatives – introduces new plants to pastures which were temporarily closed off. It has furthermore incited the nomads to create pastoral cooperatives and promised to erect shelters, and administrative buildings.

The nomad has also learned to rely on local markets, weekly or permanent, which have contributed to the integration of the local pastoral economy into the regional and national commercial fabric. The Beni Guil did not live in autarky. Their products were sold on local markets and even on foreign ones, like in Algeria, France and Great Britain. However, since 1970 it is no longer only the commercialization of their products and the purchase of merchandise that interconnects nomads to the market. The pastoral production itself, formerly relying on local resources, increasingly depends on exterior agricultural output. The local markets of Bouarfa, Tendrara and to a lesser degree Maatarka, are well provisioned and supplied. The market has thus become an important space for the diffusion of pastoral innovation, as revealed by the adoption of the sugar-beet pulp. While 48% of nomads adopted it thanks to traditional information networks (relatives: 2%, neighborhood: 10%, tribe members: 36%), 21% bought it directly from the market. The latter have asked the merchants for advice, who then showed them the nutritional qualities and the modes of preparation. Independent of the source of information, the pulp is however bought on the market. Its diffusion illustrates the speed of the adoption of external forage crops. 70% of nomads utilized it in 1991, while only 2% did so before 1970. The nomads who do not purchase pulp, bemoan its high price and the modesty of their herds. In order to face the expenses demanded by fodder, the nomads are forced to sell some of their stock on a regular basis. Camels were the first to be sacrificed. One nomad explains:

> Formerly, we did not supplement, that was good, the sheep grazed the plants of God; since they 'are eating with money', they began 'to eat themselves'.

The nomad's constraint to sell an ewe in order to feed another is voiced in the expression 'the ewe eats its own sister.' This reveals a process to which the nomad had thus far hardly been used to. So far only domestic expenses justified the sale of sheep. In the past, they said, 'the herds provided for the nomad, nowadays it is the nomad who provides for the herds.' Depending on the market, the nomad is under constraint to regularly sell a part of his herd.

Animal pests, which recently ravaged the herds, are attributed by the nomads to droughts and to the degradation of pastures. In order to care for their herds, they are increasingly forced to resort to modern medication, the distribution of which was supported by the livestock keeping service. In this regard, the most widespread innovation, adopted by 67% of nomads, is usage of parasite expellants. As is true for most innovations, a tiny minority of 2% had already used this before 1970, but its massive adoption materialized relatively late. Almost half of the users (47%) first resorted to it after 1985.

In addition to feed, the nomads are daily preoccupied to supply their livestock with water. The scarcity of water complicates this task. Besides permanent orifices like springs, wells, and the *ᶜaglat*[3], all of which are fed uniquely of phreatic sources, there exist temporary watering stations. These latter ones are constituted by swamps and underground cisterns, where waters that trickle down are accumulated. Yet rangelands are very vast and the distance between the water holes and the pastures is often large. Traditionally, the nomad used to relocate his herds towards water reservoirs. The daily march varied between 15 and 25 km, sometimes even 35 km, when water became rare due to the desiccation of temporary water holes. In 1949, the administration planned the creation of permanent watering points. They were meant to be placed at a 20 km distance from each other, each serving 10,000 ha of pasture. In 1950, for all of the high plateaus, only 120 waterspouts were counted for more than 3 million ha of steppe. This amounts to one for 30,000 ha. Moreover, 45 of them were temporary and thus empty in summer. In 1989, in the commune of Tendrara, for example, 30 deep water bores existed, of which 21 did not function. One was abandoned, six were unequipped, two were broken, seven were congested and five were dried up. Besides these boreholes there were six wells exploiting phreatic waters, one spring, and 30 underground cisterns in bad condition. In the commune of Bouarfa only four deep water bores, three springs and ten underground cisterns were counted. The steppes still lack adequate water access.

The sheep of the Beni Guil are reputed for their adaptation to the rough conditions of the steppes. Good at marching, eating little, coping comparatively well with thirst, heat and cold. In spite of these qualities the nomad is more and more forced to take care of them. After supplementary feed, the most frequently cited change for the majority of interlocutors is related to water supply. The nomads progressively abolish the habit of conducting their herds towards water holes. Herds are increasingly supplied with transported water. In 1991, 48% of nomads no longer relocated towards water orifices. Only 9% of them used transported water before 1970, compared to 56% within the 1970s and 29% in the 1980s. The nomads perceive that the life of herds progressively depends on these new practices, the transport of water and supplementary feed. One nomad emphasizes:

> The sheep no longer need to march, the water holes are distant. If the sheep walk, they will soon be scrawny and drop dead from hunger. And the one transporting water is like the one supplying complementary fodder.

In the past, the sheep used to graze while walking towards the water holes. Currently, the poverty of pasture makes long marches hazardous for the herd's health. To transport the water, a truck and two iron reservoirs are needed, one reservoir serves the water transport, the other the conservation. This kind of equipment is not in the reach of all nomads. Its price is the equivalent of 20 sheep or goats. The nomad who does not possess a means of transport is obliged to at least own a cistern, where he can conserve purchased water or water donated by a relative. If this is not the case, he must pitch his tent in the proximity of a water source, in a radius seldom exceeding two or three kilometers, which often constitutes a serious constraint to his mobility.

Since 1970, the nomad continually acquires various pieces of equipment. In addition to the use of mangers and of mobile barrels, fences (zirba) were introduced. Before the adoption of fenced paddocks the shepherd had to sleep amongst the herd. He thus had to endure the cold and rain. At night, he attached a sort of hand-held leash (lmers) to an alpha ewe. This woke him up every time the ewe tried to move away. Presenting himself to an employer with a *mers* in his hand was a way of manifesting his will and disposition to be a hard worker. The adoption of the paddock is recent. The adopters constitute 70% of the nomads, while only 2% of them possessed a paddock before 1970. For the majority (87%), the primary reason to adopt it is the protection against the wolf. The paddock also serves the separation of young animals from their mothers. This constitutes the main reason for 10% of adopters and the secondary reason, after security, for 47% of them. Finally, it has to be noted that a small number of nomads (11%) use the paddock to facilitate the distribution of supplementary feed. Those without sufficient mangers successively nourish their animals in keeping a part of their herd in paddock.

Commercialization

The nomads are conscious of the consequences that pastoral innovations induce in herd management. Until very recently, they frequented the market to sell some of their stock and to purchase basic consumer goods. Presently and to the degree that livestock producers' expenses are rising, they adopt new practices dictated by the market.

Among the Beni Guil sheep raising remains the principal activity. In 1989, they possessed 498,000 head.[4] The confederation bestowed its name upon a local sheep race, the *Beni Guil (guilliya*, also called *doughma* due to the color of its brown skin) which dominates the eastern Moroccan high plateaus. From 1930 onward, live sheep of the *Beni Guil* breed were exported to France and since 1932, its wool was also sold in England. In 1938 about 50,000 sheep, that is about 8% of the entire stock, were bought from the Beni Guil, for 110 francs per head, and were then exported to Marseille. In January and February of the following year, the same number of animals was sold at prices varying between 115 and 135 francs per head. This exceptional upsurge in the history of eastern Moroccan animal husbandry can be explained mainly by the heightened French demand for the *Beni Guil* breed.

> During the 1950s, sheep of the Beni Guil race were still appreciated for the quality of their meat, most likely due to the aromatic and salty plants of the steppes' pastures and also due to its physiognomy of small sheep with relatively well-developed lamb chops and leg-of-lamb, perfectly corresponding to the taste of French customers. Each year between 150,000 and 200,000 Beni Guil sheep (1950) are exported under the name 'mouton oranais', but they are sold at higher prices than the real sheep from the Oran region (Auriac undated, 37–41).

The nomads know that the quality of their breed stock tends to get lost by using forage crops, that the meat loses its unique taste, attributed to the consumption of aromatic plants. But they also understand that the price depends more on the animal's weight than on the taste of its meat. Every practice contributing to the augmentation of sheep weight is thus welcome. One practice, which recently spread, consists of the selective treatment of the lamb, in other words not to milk the ewes having given birth to a male. The mothers' milk is directed exclusively to the male lambs. Formerly, all ewes were milked without distinction. Presently, the selective milking is highly esteemed because of its profitability. The male lamb is pampered, for it is destined for market sale, in contrast to the female lamb, which is kept to increase the herd size. This practice is a collateral effect of the fodder supplementation. The nomads often justify it by commerce (tijara), sale (lbi^{c}), and price (taman): 'Formerly all ewes were milked, nowadays the male lamb is preferred. Today the male has become a commerce.' With supplementary feed, several nomads reduced the size of their herds: 'Ten fat sheep are better than 100 meager sheep.'

The Beni Guil also possess goats (99,000 in 1989) that do not constitute the real nomadic capital. Nomads notably raise them for their milk – for the goat is the poor man's cow. But goat meat was also cherished as was its hair, which until very recently was mixed with camel hair and used for the weaving of stripes (*flijs*) of tent canvas.

The most significant innovation regarding the composition of herds, is however the adoption of cattle. The Beni Guil did not possess cattle till the end of the nineteenth century. Even in the 1950s, cattle were still few in number. From the 1980s on, the number of cattle has risen considerably. It passed from 1,684 head in 1982 to 6,360 in 1989. Cattle raising was not adjusted to the ancient nomadic way of life. Cattle march slowly, consume lots of water and require appropriate pasture, for example, cows cannot be led to salty pastures. Thanks to the new possibilities these constraints are currently less important and the adoption of cattle reveals the profound changes of the nomadic society. Due to motorized transport, the slowness of cattle, their high water consumption, and the distance separating appropriate pastures for sheep from those for cattle, no longer constitute invincible obstacles. The truck assures the transport of cows and water, and thus enables the pastoralist to guard sheep and cattle separately. Certain nomads apply a complex system of shepherding that often demands household division and an incessant coming-and-going between distant pastures. This is worth all the efforts, for the price of calves is increasingly attractive.

In this logic of adaptation, the nomad is forced to abandon the camel and the horse. The camel no longer meets the new requirements of transportation. The nomad utilizes materials and instruments which a camel cannot transport, like cisterns, barrels and troughs. This is why the hundreds of camels that the Beni Guil still possess are no longer used for transport. They are rather confined to shepherds living in Lmangoub for an average of 10 Dh per head and month. This region, in the tribe's south, offers appropriate pastures for camels.

The horse experiences almost the same development. The nomads possessing horses were very few. 67% of them never owned a horse, 26% gave up their horses within the 1970s, and only 7% still hold horses. The horse is exhibited preeminently at riding competitions (fantasia) during familial, national or local celebrations. Whenever the pastures are inaccessible, the horse is used for scouting pastures, the inspection of the herds and the shepherd, but its pastoral function seems to be secondary.

Whatever the importance of the supply with forage crops may be, herd mobility for the search of new pastures remains the main source to feed the herd. In a setting marked by the reduction of nomadism's realm, motorized transport allows a better use of the steppes. Instead of leading the herds across degraded pastures and exhausting them even more, the herds are directly transported to adequate grazing lands. The trucks also allow the use of pastures which thus far had been inaccessible due to lacking water. In 1991, 13% of the questioned nomads possessed a truck, and 43% of them bought it after the droughts in the 1970s. Besides, 46% of the questioned nomads used trucks as means of mobility, thanks to the possibility of renting them. It can be said that almost all who are still mobile, that is 49% of the surveyed nomads, employ motorized means of transportation.

> By means of the truck it's possible to go to market and return to the tent the same day. In old times, the people endured, they only moved (in groups) by means of camels and animals. For our generation, motorized transport is disposable. So now, we are satisfied. Formerly, ten to fifteen days were required to go to the market and only little could be purchased. With motorized transport, what is available in Moroccan cities can be bought at Tendrara as well (Chadli Miloud).

But the truck is not simply some odd comfortable and fast substitute of the camel. It is not only a means of transport. Rather like the camel, the truck is at the heart of a network of social relations, albeit different social relations.

Conversion

Pastoral strategies remain largely linked to traditional habitation. The construction of a house near fields would be the first step towards conversion of the nomad. 13% of nomads questioned owned a house in the countryside. However, the conversion is only total for 6%, who have completely substituted a house for the tent. For the rest, the two kinds of habitation do not exclude each other. Possessing a house and a tent is necessary for a complex strategy, which combines agriculture and herd mobility.

Small agglomerations developed at Lmangoub, Maatarka and Laghouasil, markedly among the Oulad Youb and Oulad Mbarek. The *hayta*, metaphorically a new form of the *douar,* designates the neighborhood of a small number of houses. Those of Laghouasil are composed of fifteen household heads, who engage in agriculture. Seven of them still live in tents and eight live in houses. The oldest house was constructed in 1983. None of them has abandoned the tent. Yet, their usage of the tent is variable. It can be pitched in proximity to the house, where it morphs into some kind of extension, cooking place, or habitation for the married son. Nomads, who opted for a seasonal transhumance, pack and detach their tents according to pasture conditions. Others, who have seen their herds diminish considerably, have completely abandoned their tents, which they hope to unpack one happy day, when they reconstitute their herds.

Conserving the tent is the ultimate action that ties a nomad to the steppes. Abandoning the tent remains an extreme last option and signals complete conversion. In a society undergoing dramatic changes, compromise and multiplicity of strategies are safer. Transhumance enables the keepers of livestock to combine agricultural and pastoral activities. The owner of a house can recruit a shepherd, who will take care of the herd. Or he can entrust this task to some household members, who will move with the tent, while he and the rest of the family inhabit the house and primarily perform agricultural tasks.

Some elites, 2% of the surveyed nomads, even prefer to reside in town and to manage the activities relating to the market and public administration, while designating a part of their household to the countryside in order to look after the herd. Hence, these are households of which one part is sedentary and the other is nomadic.

The word *barra* signifying 'the exterior' has only recently been used by the Beni Guil to designate the countryside opposed to the new agglomerations of Tendrara and Bouarfa. The smaller agglomerations, such as Lmangoub and Maatarka, still are part of this 'exterior'. The nomads are aware that in a few decades the steppes have become the periphery of a dramatically expanding urban space. Thus, several nomads are no longer satisfied with life in the 'exterior': 25% have a house in town, in which 6% reside permanently. However, as it has already been noted, it is not the acquisition of a house that inaugurates the nomad's conversion. Frequently, the house, like the truck or the cistern, is a new, but indispensable accessory for the new conditions of livestock keeping. Above all, the abandonment of the tent would mark the nomad's conversion.

Settling in agglomerations does not necessarily concern poor nomads only. Among the nomads settled in town, 51% were large- or middle-scale livestock owners and possessed a double habitation (khayma and ᶜachcha). Only 19% were poor and had to be satisfied with a small tent. Sedentarization, the final conversion of a nomad, more often affects wealthy than poor pastoralists. Besides, only 4% of sedentary nomads declared that they had fixed their tents. The majority (70%) had directly transferred from the mobile habitation to the house. This indicates that immobilization of the tent is neither an intermediate stage nor necessary for sedentarization.

In town, the former nomads no longer stick to animal husbandry. The modesty of herds indicates that livestock keeping is conducted as a supplementary activity. In effect, 83% of sedentary Beni Guil have broken all ties to livestock keeping. More than half (52%) live from agriculture, 21% from urban professions, and 10% are unemployed.

Survival

After the disintegration of the *douar*, mobility, formerly a triviality, has become a heroic deed. Half of the nomads abandoned mobility after the droughts in the 1970s and 1980s. 42% of them fixed their tents during the 1970s and 36% during the 1980s. The word *mgaytan* (pl. guitan) is applied to the nomad who broke with mobility, even if his tent is fixed in the pasture areas and far from agglomerations. Still, even a settled shepherd is under constraint for reason of animal health to change his camp on average once a month. The herd must not graze, nor stay in one place for a long time. The distance between the new and old campsite rarely exceeds five kilometers. It is distinguished between *rhal* which signifies mobility and change of pastures, and *thawel* or *beddel dar*, which implies the change of campsite within the same pasture.

The majority (81%) of nomads links the immobilization of the tent to the modesty of herds. Only 7% name agricultural purposes as reason for their settlement. The abandonment of mobility prominently affects pastoralists, who own less than 60 sheep. According to these, a small herd does not justify mobility. Since the 1970s, the average herd size has been in continuous decline: it has fallen from 252 sheep to 68, and from 37 goats to 16. It must also be noted that in 1989, more than a third of nomads possessed less than 60 head. However, at the time of the *douar*, poverty did not fatalistically imply settlement. Even those with small herds could relocate thanks to the *douar* arrangement. They had the possibility of nomadizing with the grand, who in turn lent them their camels and supplied them with life's necessities. The inequality of wealth did not translate into terms of mobility and sedentarization. To the contrary, these relationships of dependence allowed the needy to stick to their nomadic life. After the dissolution of the *douar*, the poor who did not possess camels were harshly condemned to settlement or emigration. Sedentarization has been much more massive and accelerated because half of the nomads, as we have already seen, did not keep camels.

The possibility of integrating new pastoral units always exists, but requires a substantial amount of money. The poor can no longer count on the grand, who have broken with their traditional generosity. The possession of a truck demands heavy cost and thus forces its owner to lend it out for money. But for a small pastoralist, the cost of renting a truck is largely disproportionate to the herd size. The camel, in contrast to the truck, permitted the integration of the nomad and the installation of diverse social networks, some based on contracts and others on patronage. The agricultural activity rarely constitutes the primary reason for sedentarization. According to several witnesses, I assert that it is especially the immobility which favors the growing interest in agriculture, not the reverse. For example, one livestock producer explains:

> I no longer move around with my tent. Twelve or 13 years have passed, since I dug this well and now I am settled. I don't own animals for the transport of my tents, I don't have the means to rent a truck. Also, I only have few ewes and goats (20), that is why I don't move my tent anymore.

The immobility of the tent does not necessarily imply immobile herds. Some combinations are possible between traditional nomadism and sedentarization. 42% of households no longer relocate their herds, while 7% combine immobility of the tent with mobility of herds. Whenever the household disposes of a double habitation, the grand tent remains fixed, while a part of the family moves with the small tent in search of better pastures. Another pastoralist fixed his tent close to fields he cultivates, while continuing to relocate his herd:

> For three years I have not left the pasture of Sfiya. I used to move with the small tent (ᶜachcha), because the big tent (khayma) stayed fixed close to the fields. Sometimes I moved without the small tent, because it's so heavy. We conducted the herd (60 head) and we passed the night out in the open under the starry expanse. Since 1971, the grand tent has not once left the pastures of Sfiya, now it is so heavy that it no longer can be transported. Two donkeys and one old mule, what can you do with that?

A needy nomad has always had the option of pitching his tent near villages, where he could find a job. But this was almost always temporary, not irreversible. It generally was a provisional solution of pastoralists waiting for clement years. The most important center for this make-shift settlement used to be Figuig. For example, within the period from 1914 to 1936 not a single nomad was enumerated in this oasis. In 1943, 700 nomads were settled in the vicinity of Figuig. In 1948, in the aftermath of the famous drought, their number exceeded 2,100. In 1952, the year marking the end of the drought and the revival of pastoral activity, all sedentary nomads, except a few households, again removed to the pastures.

> The demographic evolution of nomads has depended on climatic conditions on the high plateaus and notably the great droughts of 1945 and the following years. The return of more favorable conditions motivated the nomads' departure with the exception of some miserable day laborers and shepherds, who together with the beggars represent the proletariat of the oasis' (Bounfous 1952).

Since the 1960s, the needy nomads, who were held by nothing to the pastures anymore, came to fix their tents or that which remained of them, close to Tendrara or Bouarfa. Presently, a group of 50 shanty tents can be observed at the periphery of Bouarfa. The new centers offer the ex-nomads the chance of new jobs. A small pastoralist, who settled six years ago on a pasture situated five kilometers from Tendrara, explains his choice:

> I don't have the means for mobility. I don't have a truck. So I moved close to the market in Tendrara. I am no longer a nomad, my herd is very modest, 24 ewes and 15 goats. I have two cows. I started raising cows after I settled. I have a chariot, but I only use it to transport water and goods bought on the *souk*.

Compared to the immobilization of the tent on a pasture, to move closer to the *souk* is the last choice in the hands of small nomads hanging on to the pastoral life. The costs of transport towards 'urban' centers can be met and the chance for family members to find jobs is greater. Yet, since the disappearance of the *douar*, the return of sedentary pastoralists to a nomadic way of life is exceptional. Sedentarization is no longer a provisional alternative in reaction to difficult and transitional climatic conditions. Sedentariness, for modest nomads, has become an integral part of their livelihood systems in the context of structural transformations limiting the mobility of people and livestock.

Conclusion

Transformations of nomadic societies are often connected to imbalances between ancestral strategies (e.g. far-distance collective mobility based on dromedary use) and new, constantly changing socio-economic contexts. In the present paper, I have tried to describe, for the Beni Guil case, empirical linkages between contextual changes – in areas such as markets, the state, etc. – and pastoral innovations. My aim was to analyze how contextual changes affect the situation of Beni Guil nomads. In order to do this, I focused on showing how the nomads integrate external pressures into their strategies. The reduction of pasture spaces was one of the major constraints that pushed the nomads to buy trucks. They were thus the very first actors to transform new opportunities into resources that serve the renewal of Moroccan nomadism. In adopting pastoral innovations, they were able to overcome the imbalances between their ancestral strategies and the challenges brought about by contextual transformation.

Notes

(1) The quantitative data presented in this article originate from a questionnaire-based survey conducted among 700 nomads of the Beni Guil tribe. The survey took place between November 1989 and June 1991 (Projet de Développement des Parcours et de l'Elevage dans l'Oriental, Ministère de l'Agriculture et de la Réforme Agraire, 1991).

(2) The nomads studied here often qualify themselves as *ᶜrab*, the alteration of the Arabic word *aᶜrāb*. The Arabic text of the Treaty of 1845 signed by Morocco and France qualified the Beni Guil as *aᶜrab*. In the French text this was improperly translated with the word "Arabs". In 1888, the tribes of the Beni Guil sent a letter to the sultan of Morocco, where the name of their confederation is preceded by the word *aᶜrab*. The elites are called *Mmaline l'aᶜrab*, the master of the nomads. The word *ᶜrab* is equally applied to all nomads in order to stress the cultural differences with sedentary groups.

(3) All the wells with little depths.

(4) In the late 19th century, the number of sheep the Beni Guil possessed was estimated at 470,500 (De La Martinière and Lacroix 1896, 273).

Bibliography and Appendices

Bibliography

Abaab, A./Bédrani, S./Bourbouze, A./Chiche, J. (1995): "Les politiques agricoles et la dynamique des systèmes agropastoraux au Maghreb", in: *Options Méditerranéennes* B 14, 139–65.

—/Elloumi, M./Genin, D./Guillaume, H./Picquet, M./Sghaier, M. (2004): *Environnement et sociétés rurales en mutation: Approches alternatives.* Paris.

Abbas, K. (2004): "Dégradation des parcours steppiques. Pour une lecture différente du rôle de l'élevage", in: *Revue Fourrages* 180, 541–56.

Abdullahi, A. M. (1990): *Pastoral Production Systems in Africa. A study of Nomadic Household Economy and Livestock Marketing in Central Somalia* (Farming Systems and Resource Economics in the Tropics 8). Kiel.

Abu-Lughod, J. L. (1980): *Rabat, Urban Apartheid in Morocco.* Princeton.

Abun-Nasr, J. (1971): *A History of the Maghrib*. Cambridge.

Ait Hamza, M. (2002): *Mobilité socio-spatiale et développement local au Sud de l'Atlas marocain (Dadès-Todgha).* (Maghreb Studien 13). Passau.

Albrow, M.: (1996) *The Global Age.* Cambridge.

Alcantara, C. H. de (ed.) (1993): *Real Markets. Social and Political Issues of Food Policy Reform.* London.

Allali, K.: Evaluation du coût des systèmes d'élevage dans la région de Ouarzazate. Projet de conservation de la diversité biologique par la transhumance dans le Haut Atlas central. Ouarzazate (unpublished report, no date).

—/Mahdi, M. (2001): "Les coopératives de la réforme agraire trente ans après", in: Bulletin Economique et Social du Maroc [BESM] 160, *Rapport du Social 2001*, 109–25.

Aloui, O./Chiche, J./Gharbaoui, A./Niamir-Fuller, M. (1998): Etude d'identification du Projet de conservation de la diversité biologique par la réhabilitation de la transhumance dans la Province de Ouarzazate. Rabat (unpublished report).

Amnesty International (2004): "Morocco/Western Sahara" 2003a, in: <*http://web.amnesty.org/web/web.nsf/print/mar-summary-eng*> (Access at Apr. 2004).

Amnesty International (2003b): "Morocco/Western: Amnesty International Delegates to Speak about Sharp Rise in Torture in Morocco", in: <*http://web.amnesty.org/library/print/ENGMDE290122003*> (Access at Apr. 2004).

Annuaire statistique du Marco = Royaume du Maroc, Ministère du Plan, Direction de la Statistique: Annuaire Statistiques du Maroc, different volumes. Rabat.

Anon (2000): "Aggravation du taux de pauvreté au Maroc Panafrican News Agency 25 October 2000", in: <*www.allafrica.com/stories/printable/ 200010250353. html*> (Access at Oct. 2000).

– (2004): "Morocco Gets US Free Trade Deal", in: *BBC World News* 23 July 2004.

Appadurai, A. (1997): *Modernity at Large. Cultural Dimensions of Globalization.* Minneapolis, London.

– (1990): "Disjuncture and Difference in the Global Cultural Economy", in: Featherstone, M. (ed.): *Global Culture. Nationalism, Globalization and Modernity.* London, 295–310.

Araba, A./Bourarache, K./Chiche, J./Yessef, M. (2002): *Les savoirs locaux dans la Province de Ouarzazate, Maroc.* Projet de conservation de la diversité biologique par la transhumance dans le Haut Atlas central. Ouarzazate (unpublished report).

Aziki, S./Bodemeyer, R. (2003): Quelques réflexions à propos des nouvelles formes d'élevage dans le sud ouest marocain: Conflit entre le nomadisme moderne et sédentaire de la région du Souss. Agadir (unpublished).

Azzimani, M. (1992): Les ressources pastorales et leur appréciation par les Ait Zekri, Province de Ouarzazate. Rabat (Mémoire de 3ème cycle en agronomie, IAV, unpublished).

Bankoff, G./Frerks, G./Hilhorst, D. (2004): *Mapping Vulnerability. Disasters, Development and People.* London, Sterling.

Barker, C./Galasiński, D. (2001): *Cultural Studies and Discourse Analysis. A Dialogue on Language and Identity.* London, Thousand Oaks, New Delhi.

Bassett, T. J./Zuéli, K. B. (2003): "The Ivorian Savanna: Global Narratives and Local Knowledge of Environmental Change", in: Bassett, T. J./Crummey, D. (eds.): *African Savannas: Global Narratives & Local Knowledges of Environmental Change.* Oxford, 115–36.

Bauman, Z.: *Liquid Modernity.* Cambridge 2000.

Beaudet, G. (1969): "Les Beni Mguild du Nord, étude géographique de l'évolution récente d'une confédération semi-nomade", in: *Révue de Géographie du Maroc* 15, 3–80.

Beck, K. (1988): "Die Kawahla von Kordofan: Ökologische und ökonomische Strategien arabischer Nomaden im Sudan", in: *Studien zur Kulturkunde* 85.

Beck, U. (2006): "Mobility and the Cosmopolitan Society", in: Bonß, W./Kesselring, S./ Vogl, G. (eds.): "Mobility and the Cosmopolitan Perspective" (Working paper series to Workshop at the Munich Reflexive Modernization Research Centre [SFB 536], 29–30 January 2004, Munich), in: <*http://www.sfb536.mwn.de/veranstaltungen/ B3_Workshop_ 0104_Dokumentation.pdf*> (Access at Sept. 2006).

Benabdellah, A. (1998): The Economics of Land Tenure and Agricultural Performance in the Mnasra Region of Morocco. St. Louis (unpublished PhD thesis).

Benabdi, A. L. (2002): La restructuration de l'espace territorial, économique et social, liée aux mutations des pratiques de l'élevage, C.R. de Timahdit, Moyen Atlas central du Maroc. Paris (Diplôme d'études approfondies, unpublished).

Benassou, L. (1998): Etude du marché ovin Dman au niveau de la vallée de Ziz (Cas: Aoufouss et Rissani). Meknès (Mémoire de 3ème cycle en agronomie, ENA, unpublished).

Bencherifa, A. (1996): "Is Sedentarization of Pastoral Nomads causing Desertification? The Case of the Beni Guil of Eastern Morocco", in: Swearingen, W./Bencherifa, A. (eds.): *The North African environment at risk.* Boulder, 117–30.

– (1988): "Agropastorale Organisationsformen im atlantischen Marokko", in: *Die Erde* 119,1, 1–15.

–/Johnson, D. (1991): "Adaptation and Intensification in the Pastoral Systems of Morocco", in: Galaty, J./Johnson, D. (eds.): *The World of Pastoralism. Herding Systems in Comparative Perspective.* New York, 394–416.

Bendaanoun, M. H. (1994): La désertification phénomène planétaire cas du Maroc: causes, impacts et moyens du lutte. Ouarzazate (ORMVAO, unpublished paper).

Benhaddou, A. (1997): *Maroc: Les élites du royaume.* Paris.

Berner, E. (1995): *Networks, Strategies and Resistance: A Critique and Reformulation of Strategic Groups Analysis.* Bielefeld.

Bernués, A./Manriqueet, E./Olaizola, A./Revilla, R. (1996): "Economic Diversity of Mountain Farms and Complementarity Strategies in Land Use", in: Zervas, N.P./Hatzimi-naoglu, J. (eds.): *The Optimal Exploitation of Marginal Mediterranean Areas by Extensive Ruminant Production Systems.* (EAAP Publications No. 83) Wageningen, 61–5.

Berque, J. (1955): *Structures Sociales de l'Haut Atlas.* Paris.

Bidwell, R. (1973): *Morocco under Colonial Rule: French Administration of Tribal Areas 1912–1956.* London.

Blench, R. (2000): *"You can't Go Home Again". Extensive Pastoral Livestock Systems: Issues and Options for the Future.* London.

Bohle, H.-G./Watts, M. (2003): "Verwundbarkeit, Sicherheit und Globalisierung", in: Gebhardt, H./Reuber, P./Wolkersdorfer, G. (eds.): *Kulturgeographie – Aktuelle Ansätze und Entwicklungen.* Heidelberg, Berlin, 67–82.

Bohle, H.-G./Gertel, J./Krings, T./Krüger, F. (1993): "Dürren und Hunger", in: Merz, B./ Plate, E. (eds.): *Naturkatastrophen & Katastrophenvorbeugung*, Weinheim, 497–512.

Bonanno, A./Busch, L./Friedland, W./Gouveia, L./Mingione, E. (eds.) (1994): *From Columbus to ConAgra. The Globalization of Agriculture and Food.* Lawrence.

Bonß, W./Kesselring, S./Vogl, G. (eds.) (2006): *Mobility and the Cosmopolitan Perspective.* (Working paper series to Workshop at the Munich Reflexive Modernization Research

Centre [SFB 536], 29–30 January 2004, Munich), in: <*http:// www.sfb536.mwn.de/ veranstaltungen/B3_Workshop_0104_Dokumentation.pdf*> (Access at Sept. 2006).

Bouderbala, N. (1999): "Les systèmes de propriété foncière au Maghreb: Le cas du Maroc", in: *Cahiers Options Méditerranéennes* 36, 47–66.

— (1997): "La modernisation et la gestion du foncier au Maroc", in: *Options Méditerranéennes* A 29, 155–64.

— (1987): "Loi et société au Maroc: Le cas des terres collectives", in: *Bulletin Economique et Social du Maroc [BESM]* 159/160/161, 63–73.

—/Chiche, J./El Aich, A. (1992): "Les terres collectives au Maroc", in: Bourbouze, A./ Rubino, R. (eds.): *Terres collectives en Méditerranée.* Villa d'Agri, 27–59.

Boudiab, A. (1981): Contribution à l'étude du système de production animale sur parcours dans la région de Ouarzazate. Rabat (Mémoire de 3ème cycle en agronomie, IAV, unpublished).

Boueiz, M./Fischer, M./Kuhn, B./Papendieck, J./Paulus, I./Stöber, S./Stumpf, H./ Ullmann, G. (1994): *Le fonctionnement du marché ovin au Maroc – Approche méthodologique et résultats de l'étude pilote au Moyen Atlas.* Berlin.

Bourbouze, A. (2001): "Pastoralisme au Maghreb: La révolution silencieuse", in: *Revue Fourrages* 161, 3–21.

— (1999): "Gestion de la mobilité et résistance des organisations pastorales des éleveurs du Haut Atlas Marocain face aux transformations du contexte pastoral maghrébin", in: Niamir-Fuller, M. (ed.): *Managing Mobility in African Rangelands.* London, 236–65.

— (1982): L'élevage dans la montagne marocaine. Organisation de l'espace et utilisation des parcours par les éleveurs du Haut Atlas. Paris-Grignon (unpublished doctoral thesis).

Bourdieu, P. (1998): *Outline of a Theory of Practice.* [Original 1972]. (Cambridge Studies in Social and Cultural Anthropology 16) Cambridge.

— (1985): *Sozialer Raum und Klassen. Leçon sur la leçon. Zwei Vorlesungen.* Frankfurt/M.

— (1984): *Distinction: A Social Critique of the Judgement of Taste.* London

— (1983): "Ökonomisches Kapital, kulturelles Kapital, soziales Kapital", in: Kreckel, R. (ed.): *Soziale Ungleichheiten.* Göttingen, 183–98.

— (1976): *Entwurf einer Theorie der Praxis, auf der ethnologischen Grundlage der kabylischen Gesellschaft.* Frankfurt/M.

Brettell, C. B./Hollifield, J. F. (eds.) (2000): *Migration Theory. Talking across Disciplines.* New York, London.

Breuer, I. (2007): *Mobilität und Existenzsicherung im ariden Marokko.* Wiesbaden (forthcoming).

– (2005): "Statistiken, oder: Wie werden 'Nomaden' in Marokko gemacht?", in: Gertel, J. (ed.): *Methoden als Aspekte der Wissenskonstruktion. Fallstudien zur Nomadismusforschung. Orientwissenschaftliche Hefte 17 = Mitteilungen des SFB "Differenz und Integration"; 8.* Halle/Saale, 55–73 (<www.nomadsed.de/publications.html#3>).

– (2004): "Mobilität und Existenzsicherung im ariden Marokko", in: Meyer, G. (ed.): *Die arabische Welt im Spiegel der Kulturgeographie*, Mainz, 256–63.

Brohman, J. (1996): *Popular Development: Rethinking the Theory and Practice of Development.* Oxford.

Burke, E. (1976): *Prelude to Protectorate in Morocco: Precolonial Protest and Resistance, 1860–1912.* Chicago.

Buttel, F. H./Busch, L./Friedland, W. H./Rudy, A. P. (eds.) (1991): *Towards a New Political Economy of Agriculture.* (Westview Special Studies in Agriculture Science and Policy) Boulder, San Francisco, Oxford.

Caidi, A. (1995): Etude comparative des systèmes d'élevage nomade, semi-nomade et sédentaire dans les parcours arides et subsahariens. Meknès (Mémoire de 3ème cycle, ENA, unpublished).

Carnegie, J./Goldman, I./Kela, E./Marumo, D./Marumo, M./Mwale, E./Ntonga, S. (2000): *Institutional Support for Sustainable Rural Livelihoods in Southern Africa: Framework and Methodology.* (Natural Resource perspectives No. 49) London.

Carney, D. (ed.) (1998): *Sustainable Rural Livelihoods: What Contribution Can We Make?* London.

Casciarri, B. (2006): "Coping with Shrinking Spaces. The Ait Unzar Pastoralists of South-Eastern Morocco", in: Chatty, D. (ed.): *Nomadic Societies in the Middle East and North Africa: Entering the 21st Century.* Leiden, 393–430.

Cass, D. Z. (2005): *The Constitutionalization of the World Trade Organization. Legitimacy, Democracy, and Community in the International Trading System.* (International Economic Law Series) Oxford.

Catusse, M. (2000): "Le débat sur les privatisations au Maroc: un procès exemplaire", in: *Monde Arabe Maghreb-Machrek* January, 38–48.

Chambers, R. (1989): "Vulnerability, Coping and Policy", in: *IDS Bulletin* 20,2, 1–7.

–/Conway, G. (1991): "Sustainable Rural Livelihoods: Practical Concepts for the 21st Century", in: *IDS Discussion Papers* 296 (1991) (<http://www.ids.ac.uk/ids>).

Chichaoui, H. (2001): Etude d'un marché régional de la viande rouge: Cas de la région de Casablanca. Rabat (Mémoire de 3ème cycle en agronomie, IAV, unpublished).

Chiche, J. (2004a): Les conflits pastoraux sur le versant sud du Haut Atlas central. Projet de conservation de la diversité biologique par la transhumance dans le Haut Atlas central. Rabat/Ouarzazate (unpublished report).

– (2004b): "L'élevage des ovins et des caprins au Maroc entre la spéculation et la rusticité", in: Dubeuf, J.-P. (ed.): L'évolution des systèmes de production ovine et caprine: avenir des systèmes extensifs face aux changements de la société. Options Méditerranéennes A 61. Zaragoza, 263–70.

– (2001): "Les effets des programmes d'encouragement à l'élevage sur la production des ovins et des caprins au Maroc", in: Rubino, R./Morand-Fehr, P. (eds.): *Production Systems and Product Quality in Sheep and Goats,* in: *Options Méditerranéennes* A 46. Zaragoza, 55–64.

– (1999): "Tendances de l'exploitation des terres et perspectives foncières, les bénéfices d'une démarche pluridisciplinaire", in: Rubino, R./Mohrand-Fehr, P. (eds.): *Systems of Sheep and Goat Production: Organization of Husbandry and Role of Extension Services = Les systèmes de production ovine et caprine : organisation de l'élevage et rôle des structures de développement*, in :*Options Méditerranéennes A 38.* Zaragoza, 311–26.

– (1998): "L'évolution des systèmes d'élevage et la gestion de l'aléa au Maroc", in: Guessous, F./Ilham, A./Rihani, N. (eds.): *Livestock Production and Climatic Uncertainty in the Mediterranean: Proceedings of the joint ANPA, EAAP, CIHEAM, FAO Symposium, Agadir, Morocco, 22–24 Oct 1998.* Wageningen, 299–319.

– (1996a): "A la recherche d'une définition des statuts fonciers au Maroc", in: *Options Méditerranéennes A* 32, 15–30.

– (1996b): "Elevage Pastoral au Maroc: prélèvement abusif ou gestion prudente des ressources", in: Zervas, N.P./Hatziminaoglu, J. (eds.): *The Optimal Exploitation of Marginal Mediterranean Areas by Extensive Ruminant Production Systems.* (EAAP Publications No. 83) Wageningen, 307–14.

– (1990): Etude socio-économique de la Commune rurale d'Itzer. Rabat (unpublished report).

Choukr-Allah, R. (1999): "Protected Culture in Morocco", in: Choukr-Allah, R. (ed.): *Protected Cultivation in the Mediterranean Region = Cultures protégées dans la région méditerranéenne.* (Cahiers Options Méditerranéennes 31) Paris, 9–15.

Claudot-Hawad, H. (2001): *Éperonner le monde: Nomadism, cosmos et politique chez les Tuaregs.* Aix-en-Provence.

Cohen, S. (2003): "Alienation and Globalization in Morocco: Addressing the Social and Political Impact of Market Integration", in: *Comparative Study of Society and History* 45, 168–89.

Coleman, J. S. (1990): *Foundations of Social Theory.* Cambridge.

Costanzo, S. (1999): "Migration aus dem Maghreb nach Italien. Soziale und räumliche Aspekte der Handlungsstrategien maghrebinischer Migranten und Migrantinnen in Kampanien, Süditalien", in: *Münchener Geographische Hefte* 80, 129–44, 160–2, 168–74.

Coupe, J. (1997): "Courting his Majesty: USAID in King Hassan's Morocco", in: *Middle East Policy* 5, 150–69.

Couvreur, G. (1968): "La vie pastorale dans le Haut Atlas Central", in: *Revue Géographique du Maroc [RGM]* 13, 3–54.

Crawford, A./Nori, M./Switzer, J. (2005): "Herding on the Brink. Towards a Global Survey of Pastoral Communities and Conflict. An occasional Working Paper from the IUCN CEESP", in: <*http://www.iisd.org/pdf/2005/security_herding_on_brink.pdf*> (Access at Sept. 2006).

Crépeau, C./Tamim, M. (1988): "Communautés pastorales et systèmes d'habitat dans le Haut Atlas de Beni-Mellal (Maroc)", in: Baduel, P. R. (ed.): *Habitat, État, Société au Maghreb.* Paris, 365–75.

Dafinger, A./Pelican, M. (2002): *Land Rights and the Politics of Integration: Pastoralists' Strategies in a Comparative View.* Halle/Saale.

Dalil, S. (2000): Etude du marché des ovins dans la région de Missour. Meknès (Mémoire de 3ème cycle, ENA, unpublished).

Daoud, Z. (1990): "Privatisations à la marocaine", in: *Monde Arabe Maghreb-Machrek* April-June, 84–101.

Darfaoui, E. M. (1998): Livestock Watering Practices in the Moroccan Pre-Sahara: Their Effects on Water and Nutrient Metabolism of Sheep in Different Body Conditions. Logan (unpublished report).

– (1995): "Evolution of Pastoral Systems in the Moroccan Pre-Sahara", in: Powell, J. (ed.): *International Rangeland Development Symposium Proceedings.* Laramie, Wyoming, 59–69.

Davis, D. K. (2007): *Forthcoming Resurrecting the Granary of Rome: Environmental History and French Colonial Expansion in North Africa.* Athens.

– (2006): "Neoliberalism, Environmentalism, and Agricultural Restructuring in Morocco", in: *The Geographical Journal* 172,2, 88–105.

– (2005a): "Indigenous Knowledge and the Desertification Debate: Problematising Expert Knowledge in North Africa", in: *Geoforum* 36, 509–24.

– (2005b): "Potential Forests: Degradation Narratives, Science, and Environmental Policy in Protectorate Morocco, 1912–1956", in: *Environmental History* 10, 211–38.

– (2004): "Desert 'wastes' of the Maghreb: Desertification Narratives in French Colonial Environmental History of North Africa", in: *Cultural Geographies* 11, 359–87.

Day, A. (2004): *Morocco: Economy in The Middle East and North Africa.* London, 841–60.

De Haan, C. (1994): "An Overview of the World Bank's Involvement in Pastoral Development", in: *Pastoral Development Network Paper* 36b, London.

De Haas, H. (2003): *Migration and Development in Southern Morocco.* Rotterdam.

Delannoy, H. (1969): "Le 'code des investissements agricoles' nouveau cadre législatif de la transformation des campagnes marocains", in: *Revue de Géographie du Maroc* 16, 151–7.

De La Martinière/Lacroix, N. (1896): *Documents pour servir à l'étude du Nord Ouest Africain.* Gouvernement general de l'Algérie, vol. II.

De Monts de Savasse, R. (1951): *L'élevage dans une Tribu Berbère de Montagne du Sud Marocain, Les Ait Atta du Sahara.* (Reports du CHEAM No. 1881) Paris.

Denoeux, G. P./Maghraoui, A. (1998): "The Political Economy of Structural Adjustment in Morocco", in Layachi, A. (ed.): *Economic Crisis and Political Change in North Africa.* Westport, CT 55–87.

DFG (1993) <Deutsche Forschungsgemeinschaft>: *Naturkatastrophen und Katastrophenvorbeugung. Bericht zur IDNDR.* Weinheim.

DFID (2001) <Department for International Development> : "Notes d'information sur les moyens d'existence durables", in: *<http://wwwlivelihoods.org/info/ guidance_sheets_rtfs/>* (Access at Feb. 2004).

— (2000): *Sustainable Livelihoods Guidance Sheets.* London (<www.dfid.gov.uk>).

Diallo, Y./Schlee, G. (eds.) (2000): *L'éthnicité peule dans des contexts nouveaux. La dynamique des frontières.* Paris.

Dillman, B. (2001): "Facing the Market in North Africa", in: *The Middle East Journal* 55, 198–215.

Djoudi, H. (2001): "Un système d'élevage, multiples gestions: comparaison des performances de production de trois sous systèmes d'élevage ovin pratiqués dans le contexte agro-sylvo-pastoral du Moyen Atlas", in: Boulanouar, B./Matthess-Guerrero, A. (eds.): *Rapport annuel du Programme Viandes Rouges 2000.* Rabat, 100–8.

— (1998): Situation actuelle et fonctionnement de l'élevage ovin chez les Ireklaouen au Moyen Atlas. Approche méthodologique et résultats préliminaires sur trois sous systèmes d'élevage. Meknès (unpublished).

— (1997): "Contribution à l'étude de la production en phytomasse arbustive des parcours des Ireklaouen, par le calcul de régression entre le phytovolume et la phytomasse", in: Boulanouar, B./Matthess-Guerrero, A./Morkramer, G. (eds.): *Rapport annuel du Programme Viandes Rouges 1997.* Rabat, 110–5.

— (1996): Conduite actuelle des troupeaux ovins sur parcours et en forêt dans le Moyen Atlas. Phase pilote méthodologique Ireklaouen. Meknès (unpublished).

—/Hoffmann, I./Steinbach, J. (2000): "Identifizierung pastoraler Systeme mittels partizipativer Methoden im Mittleren Atlas", in: *International Agricultural Research. A Contribution to Crisis Prevention.* (Proceedings of Deutscher Tropentag, Hohenheim, 11–12 October 2000).

—/Kasten, I. (1998): Catalogue d'identification par les items des espèces pastorales sur les parcours collectifs des Ireklaouen en vue de l'étude du comportement alimentaire des ovins par la coprologie. Rabat (unpublished).

Douidich, M.: "The Role Played by Households in Growth: Case Study of Morocco", in: *<http://www.gdnet.org/pdf2/gdn_library/global_research_projects/explaining_growth/Morocco_households_final.pdf>* (Access at Sept. 2006).

Dunn, R. E. (1977): *Resistance in the Desert: Moroccan Responses to French Imperialism 1881–1912.* London.

DPA (2002): Enquête rurale participative dans la zone du Projet de développement de l'élevage et des parcours de l'Oriental (Maroc). Rome (unpublished).

DPA-Bouarfa (2002): Rapport d'achèvement du projet de développement des parcours et de l'élevage dans l'oriental. Bouarfa (unpublished report).

Dresch, J. (1941): *Document sur les genres de vie de montagne dans le massif central du Grand Atlas.* Tours.

— (1939): "Les genres de vie de la montagne dans le massif du Toubkal", in: *Revue Géographique du Maroc [RGM] 1.*

Driouchi, A./Ait El Mekki, A./El Kasmi, H. (1995): "Production et commercialisation des animaux et produits d'origine animale au Maroc", in: *Options Méditerranéennes* B 14, 192–218.

Düvell, F./Jordan, B. (2002): *Irregular Migration. The Dilemmas of Transnational Mobility.* Cheltenham, Northampton.

EIU (2005) < Economist Intelligence Unit>: *Morocco: Country Profile 2005.* London.

— (2003): *Morocco: Country Profile 2003.* London.

— (2000): *Morocco: Country Profile 1999–2000.* London.

El Alaoui, M. (2002): Le statut juridique des terres collectives au Maroc et les institutions coutumières et locales dans la zone du projet CBTHA. Projet de conservation de la diversité biologique par la transhumance dans le Haut Atlas central. Ouarzazate (unpublished report).

— (1996): "Les coopératives pastorales 'ethno-lignagères' du Maroc oriental", in: *Options Méditerranéennes* A 32, 129–39.

El Amari, A. (1986): *Tarikh Addu'ayyif.* Rabat.

El Asri, N. (1999): "Agriculture: les temps modernes du monde rural arrivent", in: *Le Matin du Sahara et du Maghreb* 17 January 1999, (<http://lematin.press.ma/lejournal/1999/01/17/eco1.htm>, Access at Oct. 2000).

El Garmai, A. (1989): Contribution à l'étude de l'élevage dans la Commune rurale d'Ain Beni Mathar (Maroc oriental). Rabat (Mémoire de 3ème cycle, IAV, unpublished).

El Gharbaoui, A. (1985): "Amélioration des terrains de parcours", in: *Hommes, Terre et Eaux: Revue Marocaine des Sciences Agronomiques et Vétérinaires* 16, 59–64.

El Harizi, K. (1998): *Morocco Case Study: Blending Old with New Institutions – An Innovative Approach to Sustainable Range Land Development and Management in a Traditional Society.* Washington, DC.

El Khyari, T. (1987): *Agriculture au Maroc.* Morocco.

El Meskini, M. (1997): Utilisation et gestion des ressources pastorales et caractérisation des systèmes de production dans les provinces de Guelmim et Assa-Zag. Rabat (Mémoire de fin d'études, IAV, unpublished).

Ellis, F. (2000): *Rural Livelihoods and Diversity in Developing Countries.* Oxford.

ENNVM (1999): Enquête nationale sur les niveaux de vie [National survey on living standards]. Rabat, Direction de la statistique.

Etienne, B./Karsenty, J. C./Lazarev, G./Ollivier, M./Poncet, J. (1977): *Problèmes agraires au Maghreb.* Paris.

Evers, H.-D. (1987): "Subsistenzproduktion, Markt und Staat: Der sogenannte Bielefelder Verflechtungsansatz", in: *Geographische Rundschau* 39, 136–40.

—/Schiel, T. (1988): *Strategische Gruppen: Vergleichende Studien zu Staat, Staatsbürokratie und Klassenbildung in der Dritten Welt.* Berlin.

Fairhead, J./Leach, M. (1998): *Reframing Deforestation: Global Analyses and Local Realities: Studies in West Africa.* London.

— (1996): *Misreading the African Landscape: Society and Ecology in a Forest-Savanna Mosaic.* Cambridge.

Faist, T. (2000): *The Volume and Dynamics of International Migration and Transnational Social Spaces.* Oxford.

FAO (2001) <Food Agricultural Organization>: *Pastoralism in the New Millennium.* (FAO Animal Production and Health Paper 150). Rome.

Faouzi, M. (2002): "Formation continue au profit du personnel communal", in: *Le Matin du Sahara et du Maghreb* 11. (<www.lematin.press.ma/journal/printarticle.asp?tb=article20027ID=15651>, Access at July 2003).

Favell, A. (2001): "Migration, Mobility and Globaloney: Metaphors and Rhetoric in the Sociology of Globalization", in: *Global Networks* 1,4, 389–98.

Fay, G. (1987): "Ré-inventer la jma^{c}a. Réflexions pour le développement des milieux pastoraux", in: *Bulletin Economique et Social du Maroc [BESM]* 159/160/161, 63–73.

— (1986): "Désagrégation des collectivités et dégradation des milieux dans le Haut Atlas Marocain", in: *Revue de l'Occident Musulman et de la Méditerrané* 41/42, 234–48.

Fine, B./Heasman, M./Wright, J. (1996): *Consumption in the Age of Affluence. The World of Food.* London, New York.

Finke, P. (2005): *Nomaden im Transformationsprozess. Kasachen in der post-sozialistischen Mongolei.* (Kölner Ethnologische Studien Band 29) Münster.

Fold, N./Pitchard, B. (eds.) (2005): *Cross-Continental Food Chains.* London, New York.

Food First (2002): "Negative Impacts of World Bank Policies", in: *News and Views* Spring, 1–4.

Friedman, K. (1984): "Households as Income-Pooling Units", in: Evers, H.-D./Smith, J./ Wallerstein, I. M. (eds.): *Households and the World-Economy.* Beverly Hills, London, 37–55.

GAIN (2003) <Global Agriculture Information Network>: *Morocco. Trade Policy Monitoring. Free Trade Negotiations with EU: Tomatoes vs Bread Wheat.* (GAIN Report MO3024).

Galaty, J. G. (2004a): *Modern Mobility: Transformations in Pastoralist Land Tenure and Territoriality in Africa.* (Paper presented to the workshop: "Conditions of Pastoral Mobility". Max Planck Institute for Social Anthropology, June 17–19 2004) Halle/Saale.

— (2004b): *Time, Terror, and Pastoral Inertia: Sedentarization and Conflict in Northern Kenya.* (Paper presented to the workshop: "Conditions of Pastoral Mobility", Max Planck Institute for Social Anthropology, June 17–19 2004), Halle/Saale.

Gallie, G. (1953): *Introduction à l'étude de la vie économique et sociale des Hauts Plateaux du Maroc oriental.* Paris.

Garrigues-Cresswell, M. (1985): "Pasteurs-agriculteurs du Haut Atlas occidental", in: *Production pastorale et société* 16, 5–31.

Gellner, E. (1969): *Saints of the Atlas.* London.

Gerlings, J. H./Jongmans, D. G. (1956): *Les Ait Atta: Leur sédentarisation.* Amsterdam.

Gertel, J. (ed.) (2005a): *Methoden als Aspekte der Wissenskonstruktion – Fallstudien zur Nomadismusforschung. Orientwissenschaftliche Hefte 17 = Mitteilungen des SFB "Differenz und Integration"; 8.* Halle/Saale (<www.nomadsed.de/publications.html#3>).

— (2005b): "Inscribed Bodies within Commodity Chains: Global Wheat and Local Insecurity", in: Fold, N./Pitchard, B. (eds.): *Cross-Continental Food Chains.* London, New York, 109–23.

— (2002): "Globalisierung, Entankerung und Mobilität: Analytische Perspektiven einer gegenwartsbezogenen geographischen Nomadismusforschung", in: Leder, S./Streck, B. (eds.): *Nomadismus aus der Perspektive der Begrifflichkeit. Orientwissenschaftliche Hefte 3 = Mitteilungen des SFB „Differenz und Integration" 1.* Halle/Saale, 57–88 (<www.nomadsed.de/publications.html#3>).

– (1997): *Kairo. Nahrungssicherheit und Verwundbarkeit in Kontext der Globalisierung.* Habilitationsschrift, Freiburg.

Giddens, A. (1990): *The Consequences of Modernity.* Stanford.

– (1995): *The Constitution of Society. Outline of the Theory of Structuration.* Cambidge, Oxford.

Goldman, I./Carnegie, J./Marumo, M./Marumo, D./Kela, E./Ntonga, S./Mwale, E. (2000): "Institutional Support for Sustainable Rural Livelihoods in Southern Africa: Framework and Methodology", in: *Natural Resource perspectives no 49. Overseas Development Institute (ODI), Department for International Development (DFID).*

Goldman, M. (2001): "Constructing an Environmental State: Eco-Governmentality and other Practices of a 'Green' World Bank", in: *Social Problems* 48, 499–523.

González de la Rocha, M. (2006): "Vanishing Assets: Cumulative Disadvantage among the Urban Poor", in: *The ANNALS of the American Academy of Political and Social Science [AAPSS]* 606,1, 68–94.

Goodman, D./Watts, M. J.(eds.) (1997): *Globalising Food. Agrarian Questions and Global Restructuring.* London, New York.

Gow, D./Burzlaff, D./Duggan, W./Martin, R. (1985a): *Opportunities for AID in the Extensive Livestock Sector of Morocco: Executive Summary.* Washington, D.C.

– (1985b): *The Range Management Improvement Project in Morocco: An Evaluation.* Washington, D.C.

Guessous, F. (1991): *Productions fourragères et systèmes animaux.* (Editions Actes 118) Rabat.

–/Kabbali, A./Narjisse, H. (1992): "Utilization des chaumes des céréales par les ruminants", in: Guessous, F./Kabbali, A./Narjisse, H. (eds.): *Livestock in the Mediterranean Cereal Production Systems.* (EAAP Publications No. 49) Wageningen, 82–8.

Guiraudon, V./Joppke, Ch. (eds.) (2001): *Controlling a New Migration World.* London, New York.

Habib, A. (1982): *Lutte contre l'ensablement des palmeraies dans le Sud Marocain.* Rome.

Hadji, Z./Sekkate, M. R. (2000): "Sauvegarde et protection du cheptel en période de sécheresse. Cas du Maroc", in: Guessous, F./Ilham, A./Rihani, N. (eds.): *Livestock Production and Climatic Uncertainty in the Mediterranean: Proceedings of the joint ANPA, EAAP, CIHEAM, FAO Symposium, Agadir, Morocco, 22–24 Oct 1998.* Wageningen, 27–34.

Hall, St. (ed.) (2002): *Represantation. Cultural Representations and Signifying Practices.* London, Thousand Oaks, New Delhi.

Hammoudi, A. (1977): La vallée de l'Azzaden. Contribution à la sociologie du Haut-Atlas marocain. Paris (unpublished doctoral thesis).

—/Mahmoudi, M./Rachik, H./Tozy, M. (1992): Etude sur le changement, l'innovation et les attitudes, PDPEO, phase II. Rabat (unpublished report).

Hammoudou, M. (1993): L'exploitation des terres à pâturage dans la moyenne vallée du Drâa. Projet de lutte contre la désertification dans la vallée du Drâa. Ouarzazate (unpublished report).

— (1989): Qualité des terres à pâturage. Essai de mise au point d'une méthode d'étude comparative de la perception des éleveurs et de l'analyse scientifique dans la région de Tounfite, Haut Atlas Oriental. Rabat (Mémoire de 3ème cycle en agronomie, IAV, unpublished).

Hardin, G. (1974): "Living on a lifeboat", in: *BioScience* 24,10, 561–8.

— (1968): "The Tragedy of the Commons", in: *Science* 162, 1243–8.

Harrach, M. (1997): Organisation coopérative de gestion de l'espace et dynamique du système agro-pastoral dans l'oriental du Maroc. Cas de la province de Figuig. Montpellier (unpublished master of science thesis).

Harris, N. (1995): *The New Untouchables. Immigration and the New World Worker.* London, New York.

Harriss, B./Gillespie, S./Pryer, J. (1990): "Poverty and Malnutrition at Extremes of South Asian Food Systems", in: *Economic and Political Weekly*, December 22, 2783–99.

Harriss-White, B.: "Introduction: Visible Hands", in: Harriss-White, B. (ed.): *Agricultural Markets from Theory to Practice*. Macmillan 1999, 1–36.

Hart, D. M. (1984): *The Ait Atta of Southern Morocco. Daily Life and Recent History.* Cambridge.

— (1981): *Dadda Atta and his Forty Grandsons. The Socio-Political Organization of the Ait Atta of Southern Morocco.* Cambridge.

— (1978): "Les institutions des Ait Morhrad et Ait Hadiddou", in: *Bulletin Economique et Social du Maroc [BESM]* 138/139, 57–83.

Harvey, D. (2004): "The 'new' Imperialism: Accumulation by Dispossession", in: *The Socialist Register*, 63–87.

Hatt, D. G. (1996): "Establishing Tradition: the Development of Chiefly Authority in the Western High Atlas Mountains of Morocco, 1890–1990", in: *Journal of Legal Pluralism and Unofficial Law* 37/38, 123–50.

HCP (2005) <Royaume du Maroc, Haut Commissariat du Plan>: Pauvreté, développement humain et développement social au Maroc. Rabat.

HCP (2001) <Haut Commissariat au Plan>: *National Survey of Consumption and Household Expenditures 2000/2001, Rapport de synthèse.* Rabat.

Hendry, J. (2000): "Strategic Decision Making, Discourse, and Strategy as Social Practice", in: *Journal of Management Studies* 37,7, 955–77.

Herzenni, A. (1993): "Gestion des ressources et conditions du développement local dans la Haute Montagne de la province d'Azilal", in: Bencherifa, A. (ed.): *Utilisation et conservation des ressources: Montagnes et Hauts-Pays de l'Afrique 2.* Rabat, 333–45.

Hirschmann, A. O. (1970): *Exit, Voice, and Loyalty. Responses to Decline in Firms, Organizations and States.* Cambridge.

Hsain, A. (1989): Le fonctionnement du Conseil Communal et son rôle. Cas de la C.R. d'Ireklaouen. Rabat (Mémoire de 3ème cycle en agronomie, IAV, unpublished).

Hughes, A./Reimer, S. (eds.) (2004): *Geographies of Commodity Chains.* London.

Idil, A. (2000): "L'évolution du système agro-pastoral dans le Moyen Atlas du Nord-Est (Maroc)", in: *Espace Géographique et Société Marocaine*, 5–35.

Jaafar, B./Ramdan, A./Yessef, M. (1996): "Le partage des terres collectives dans la moyenne vallée du Dra (Maroc): atouts et contraintes pour la réhabilitation des parcours", in: *Option Méditerranéennes* B 14, 169–76.

Jacques-Meunié, D. (1982): *Le Maroc Saharien des origines à 1670.* Paris.

Janzen, J. (2004): *Changing Patterns of Pastoral Mobility in Mongolia.* (Paper presented at the Workshop "Conditions of Pastoral Mobility", June 17–19, 2004, Max Planck Institute for Social Anthropology) Halle/Saale.

– (1999): "Einführende Gedanken zum Thema 'Räumliche Mobilität und Existenzsicherung'", in: Janzen, J. (ed.): *Räumliche Mobilität und Existenzsicherung: Fred Scholz zum 60. Geburtstag.* (Abhandlungen – Anthropogeographie, 60), Berlin.

Jardak, M. (1996): *La* sédentarisation des nomades dans les steppes arides, dynamique des exploitations familiales (cas des Ait Isfoul, Anagam, Province de Ouarzazate). Rabat (Mémoire de 3ème cycle, IAV, unpublished).

Jarir, M. (1989): "Vers une disparition rapide du nomadisme au Sahara Marocain (1): Le cas du Dra Moyen", in: URBAMA (ed.): *Le Nomade, l'Oasis et la Ville.* (Fascicule de Recherche No. 20) Tours, 63–72.

Johnson, C. A. (1997): "Rules, Norms and the Pursuit of Sustainable Livelihoods" (IDS Working Paper 52), in: *<http://www.ids.ac.uk/ids/bookshop/wp/ wp52.pdf>* (Access at Sept. 2006).

Kalumenos-Auf der Mauer, N. (1994): *Zwei Dörfer der marokkanischen Präsahara im wirtschaftlichen Wandel: Taloust und Ait Aissa.* Bonn.

Kamil, H. (1999): Pour une problématique d'aménagement pastoral, esquisse d'éléments d'analyse socio-anthropologique du développement. Cas de Missour. Montpellier (unpublished master thesis).

– (1997): "De la modernisation dans les zones arides marocaines. Esquisse pour une approche anthropologique", in: *Options Méditerranéennes* A 29, 235–42.

– (1993): Penser et agir 'jdoub'. Gestion du risque et rationalité du comportement d'élevage des pasteurs nomades de Missour. Aix-en-Provence (Mémoire de DEA, unpublished).

Kassambara, I. (1984): Etude des systèmes d'utilisation des ressources végétales chez les Illemchenes (Ait Atta) du Sarhro. Rabat (Mémoire de 3ème cycle, IAV, unpublished).

Kaufmann, V. (2004): "Motility: a Key Notion to Analyse the Social Structure of Second Modernity?", in: Bonß, W./Kesselring, S./Vogl, G. (eds.): "Mobility and the Cosmopolitan Perspective" (Working paper series to Workshop at the Munich Reflexive Modernization Research Centre [SFB 536], 29–30 January 2004, Munich), in: <http://www.sfb536.mwn.de/veranstaltungen/B3_Workshop_0104_Dokumentation.pdf> (Access at Sept. 2006).

Kaufmann, V./Bergman, M. M./Joye, D. (2004): "Motility: Mobility as Capital", in: *International Journal of Urban and Regional Research* 28,4, 745–56.

Khalil, M. (1997): La filière de la viande ovine sur les hauts plateaux de l'Oriental au Maroc. L'analyse d'un système complexe et performant. Montpellier (unpublished master of science thesis).

Kharoufi, M. (1984): "Sociétés pastorales en crise au Maroc: Le cas des 'moutonniers' Beni Meskine d'El Bourouj après la sécheresse", in: *Horizons maghrébins / Droit à la mémoire* 1, 29–50.

Kihlström, A./Israel, J. (2002): "Communicative or strategic action – an examination of fundamental issues in the theory of communicative action", in: *International Journal of Social Welfare* 11, 210–8.

Kimbrell, A. (ed.) (2002): *The Fatal Harvest Reader. The Tragedy of Industrial Agriculture.* Washington, Covelo, London.

Korachi, M. T. B. (1995): "Aperçu général sur la lutte contre la désertification au Maroc", in: *Hommes, Terre et Eaux: Revue Marocaine des Sciences Agronomiques et Vétérinaires* 25, 27–35.

Fläming, T./Korf, B./Schenk, C./Ziebell, M./Ziegler, J. (2001): *Conflict – Threat or Opportunity? Land Use and Coping Strategies of War-affected Communities in Trincomalee, Sri Lanka.* (Schriftenreihe des Seminars für Ländliche Entwicklung 195) Berlin.

Kraus, W. (1991): *Die Ayt Hdiddu. Wirtschaft und Gesellschaft im Zentralen Hohen Atlas.* Wien.

Krings, T./Müller, B. (2001): "Politische Ökologie: Theoretische Leitlinien und aktuelle Forschungsfelder", in: *Heidelberger Geographische Arbeiten* 112, 93–116.

Le Gall, F. (1995): Le développement des zones de parcours. Situation actuelle, bilan des actions de développement, éléments de stratégie. Washington (unpublished report).

Le Houerou, H. N. (1993): "Environmental Aspects of Fodder Trees and Shrubs Plantation in the Mediterranean Basin", in: Papanastasis, V. (ed.): *Fodder Trees and Shrubs in the Mediterranean Production Systems.* Luxemburg, 11–34.

Leach, M./Mearns, R. (eds.) (1996): *The Lie of the land: Challenging Received Wisdom on the African Environment.* London.

Le Journal Hebdomadaire (2005): "Dossier: Sahara: Les germes de la révolution", in: *Le Journal Hebdomadaire* No. 206.

Lecestre-Rollier, B. (1992): Anthropologie d'un espace montagnard: les Ait Bouguemez du Haut Atlas marocain. Paris (unpublished doctoral thesis).

Leder, St./Streck, B. (eds.) (2005): *Shifts and Drifts in Nomad-Sedentary Relations.* (Nomaden und Sesshafte 2) Wiesbaden.

Lefébure, C. (1987a): "Ait-Khabbach du Sud-Est pré-saharien. Réorienter le vouloir vivre collectif", in: *Bulletin Economique et Social du Maroc [BESM]* 159/160/161, 23–41.

— (1987b): "Ayt Khebbach, impasse sud-est. L'involution d'une tribu marocaine exclue du Sahara", in: *Revue de l'Occident Musulman et de la Méditerranée* 41–2, 136–57.

— (1986): "Accès aux ressources collectives et structure sociale: L'estivage chez les Ayt Atta", in: *Production pastorale et société* 18, 113–26.

Little, P. D./Watts, M. J. (eds.) (1994): *Living under Contract. Contract Farming and Agrarian Transformation in Sub-Saharan Africa.* Wisconsin.

Loseby, M./Piccinini, A. (eds.) (2001): *Agricultural Policies in Europe and the USA. Farmers between Subsidies and the Market.* New York.

Mackintosh, M. (1990): "Abstract Markets and Real Needs." In: Bernstein, H./Crow, B./ Mackintosh, M. (eds.): *The Food Question. Profits versus People?* London, 43–53.

MADR (2001) <Ministère de l'Agriculture, du Développement Rural et des Eaux et Forêts> (ed.): *Programme d'Action National de Lutte contre la Désertification.* Salé.

— (2000): *Colloque national de l'agriculture et du développement rural.* Rabat.

Mahamat, A. K. (1993): Etude de la commercialisation des ovins et des caprins dans les régions de Chichaoua et de Rehamna, Intégration au Marché. Rabat (Mémoire de 3ème cycle en agronomie, IAV, unpublished).

Mahdi, M. (2005): "Anthropologie et demande sociale: à propos des communautés de pasteurs", in: *Prologues* 32.

— (ed.) (2002a): *Mutations sociales et réorganisation de espaces steppiques.* Casablanca.

— (2002b): "Du nomadisme aux nouvelles formes d'élevage", in: Mahdi, M. (ed): *Mutations sociales et réorganisation des espaces steppiques.* Casablanca, 65–90.

— (1999): *Pasteur de l'Atlas: Production pastorale, droit et rituel.* Casablanca.

Mahmoudi, A. (1996): La transformation d'un système agro-sylvo-pastoral confronté aux changements écologiques, économiques et sociaux de la montagne marocaine. Montpellier (unpublished master of science thesis).

MAMVA (1998) <Ministère de l'Agriculture et de Mise en Valeur Agricole>: Parc naturel d'Ifrane. Plan directeur d'aménagement et de gestion, volume I. Rabat (unpublished).

– (1996): *Bilan de la campagne agricole 1995/1996.* Rabat.

Manriqueet, E./Olaizola, A./Bernués, A./Revilla, R. (1996): "Ministère de l'Agriculture et de la Mise en Valeur Agricole, Economic diversity of mountain farms and complementarity strategies in land use", in: Zefas, N. P./Hatziminaoglo, J. (eds.): *The optimal exploitation of marginal Mediterranean areas by extensive ruminant production systems.* EAAP Publications No 83, 61–5.

Maroc (Direction de la Statistique) (2001a): *Annuaire statistique du Maroc.* Rabat.

– (Ministère de l'Agriculture, du Développement Rural et des Eaux et Forêts) (2001b): *Programme d'action national de lutte contre la désertification.* Rabat.

– (Secrétariat d'Etat Chargé de l'Environnement) (2001c): *Rapport sur l'état de l'environnement du Maroc.* Rabat.

– (Ministère de la Prévision Economique et du Plan) (2000): *Le plan développement économique et social 2000–2004.* Rabat.

– (Ministère de la Prévision Économique et du Plan) (1998): *Note d'orientation pour la préparation du plan quinquennal 1999–2003.* Rabat.

– (Dahir no. 1–95–10 du ramadan 1415, 22 février 1995) (1995): "portant promulgation de la loi no. 33–94 relative aux périmètres de mise en valeur en bour", in: *Bulletin Officiel du Royaume du Maroc.* Rabat, 391–7.

Marrakchi, B. L. (1993): Estimations économétriques des offres et des demandes relatives aux marchés des viandes bovins et ovins au Maroc: Cas de Marrakech, Rabat et Casablanca. Meknès (Mémoire de 3ème cycle en agronomie, ENA, unpublished).

Maurer, G. (1996): "L'homme et les montagnes atlasiques au Maghreb", in: *Annales Géographiques* 587, 47–72.

McAfee, K. (1999): "Selling Nature to Save It? Biodiversity and Green Developmentalism", in: *Society and Space* 17, 133–54.

McCarthy, J./Prudham, S. (2004): "Neoliberal Nature and the Nature of Neoliberalism", in: *Geoforum* 35, 275–83.

McCay, B. J./Acheson, J. A. (eds.) (1987): *The Question of the Commons: The Culture and Ecology of Communal Resources.* Tucson.

McMicheal, P. (ed.) (1995): *Food and Agrarian Orders in the World-Economy.* Westport, London.

Meadows, N./Morton, J. (2000): "Pastoralism and Sustainable Livelihoods: An Emerging Agenda" (Policy Series 11), Greenwich, in: *http://www.nri.org/publications/policyseries/PolicySeriesNo11.pdf* (Access at Sept. 2006).

Mellaoui, F. (2000): *Nomadische, bäuerliche und staatliche Waldnutzungssysteme Nordafrikas im Widerstreit: Probleme und Perspektiven an Fallbeispielen aus Marokko.* Berlin.

Michmerhuizen, J. S. (1999): Salinity Effects on Argan (Argania spinosia (L) Skeels) Germination and Juvenile Growth. East Lansing (unpublished master thesis).

Milanovic, B. (2002): "True World Income Distribution, 1988 and 1993: First Calculation Based on Household Survey Alone", in: *The Economic Journal* 112, 51–92.

Miller, J. (1996): "Sustained Past and Risky Present: The Tafilalt Oasis of Southeastern Morocco", in: Swearingen, W./Bencherifa, A. (eds.): *The North African Environment at Risk.* Boulder, 55–69.

Mol, A./Law, J. (1994): "Regions, networks and Fluids: Anaemia and Social Topology", in: *Social Studies of Science* 24, 641–71.

Montagne, R. (1989): *Les Berbères et le Makhzen dans le Sud du Maroc.* Paris.

Moore, P. W. (2003): "The Newest Jordan: Free Trade, Peace and an Ace in the Hole", in: *Middle East Report Online* June (<www.merip.org/mero/mero062603.html>, Access at March 2004).

Mountasser, M. (1987): Collectivités traditionelles et espaces ruraux montagnards dans les zones de l'arrière-pays atlasiques méridionaux. Le cas des Ayt Seddrate du Dadss. Aix-en-Provence (unpublished doctoral thesis).

Müller-Hohenstein, K. (1978): *Die ostmarokkanischen Hochplateaus. Ein Beitrag zur Regionalforschung und zur Biogeographie eines nordafrikanischen Trockensteppenraumes.* (Erlanger Geographische Arbeiten Sonderband 7) Erlangen.

—/Popp, H. (1990): *Marokko. Ein islamisches Entwicklungsland mit kolonialer Vergangenheit.* Stuttgart.

Msika, B./El Harizi, K./Bourbouze, A./Lazarev, G. (1997): *Projet de développement de l'élevage et des parcours de l'Oriental.* Rome.

Nathan Associates (2003): *Assessment of Morocco's Technical Assistance Needs in Negotiating and Implementing a Free Trade Agreement with the United States.* Arlington.

Neumann, R. P. (1998): *Imposing Wilderness: Struggles over Livelihood and Nature Preservation in Africa.* Berkeley, Los Angeles.

Niamir-Fuller, M. (ed.) (1999): *Managing Mobility in African Rangelands. The Legitimization of Transhumance.* London.

Niclausse, M. (1956): *Une Tribu du Sud Marocain, Les Ait Atta du Sahara.* (Reports du CHEAM No. 2681) Paris.

– (1954): *Rapports entre Nomades et Sédentaires dans le Couse du Dra.* (Reports du CHEAM No. 2306) Paris.

Norton, B. (1995): Méthode quantitative et qualitative de l'étude de la production des parcours. Rapport de Mission. Meknès (unpublished report).

Nouiga, D. (2003): *Le cri des nomades.* Paris.

ORMVAO <Office Régional de Mise in Valeur Agricole> (1993): Projet pilote de mise en valeur des terrains de parcours de la région de Ouarzazate. Ouarzazate (unpublished).

– (1992): Projet de développement pastoral des terrains de parcours de Ouarzazate. Ouarzazate (unpublished).

– (1990): Stratégie de développement des terrains de parcours dans la zone d'action de l'ORVMA de Ouarzazate. Ouarzazate (unpublished).

– (1983): Évaluation des phénomènes de désertification dans la province de Ouarzazate. Ouarzazate (unpublished).

Papanastasis, V. (1996): "Selection and Utilization of Fodder Trees and Shrubs in Mediterranean Extensive Livestock Production Systems", in: Zervas, N. P./Hatziminaoglu, J. (eds.): *The Optimal Exploitation of Marginal Mediterranean Areas by Extensive Ruminant Production Systems.* (EAAP Publications No. 83) Wageningen, 251–61.

Pardini, A. (2004): "Mediterranean Pastoral Systems and the Threat of Globalization", in: Ferchichi, A. (comp.), *Réhabilitation des pâturages et des parcours en milieux méditerranéens = Rangeland and Pasture Rehabilitation in Mediterranean Areas,* in: *Cahiers Options Méditerranéennes 62.* Zaragoza, 155–68.

Pascon, P. (1980): *Etudes Rurales: idées et enquêtes sur la campagne marocaine.* Rabat.

– (1977): *Le Haouz de Marrakech. Histoire sociale et structures agraires.* Rabat.

Paulus, I./Boueiz, M./Fischer, M./Kuhn, B./Papendieck, J./Stöber, S./Stumpf, H./Ullmann, G. (1994): *Le fonctionnement du marché ovin au Maroc – Approche méthodologique et résultats de l'étude pilote au Moyen Atlas.* Berlin.

Payne, R. (1986): "Food Deficits and Political Legitimacy: The Case of Morocco", in: Commins, S./Lofchie, M./Payne, R. (eds.): *Africa's Agrarian Crisis: Roots of Famine.* Boulder, 153–72.

Peet, R. (1999): *Theories of Development.* New York .

Peltier, J.-P. (1982): La végétation du bassin versant de l'Oued Souss (Maroc). Grenoble (unpublished doctoral thesis).

Peluso, N. L./Ribot, J. C. (2003): "A Theory of Access", in: *Rural Sociology* 68,2, 153–81.

Pennell, C. R. (2000): *Morocco since 1830: a History.* New York.

Pfeifer, K. (1999): "How Tunisia, Morocco, Jordan and even Egypt Became IMF 'Success Stories' in the 1990s", in: *Middle East Report* Spring, 23–7.

Popp, H. (1983): *Moderne Bewässerungslandwirtschaft in Marokko.* Erlangen.

Pribram, K. (1983): *A History of Economic Reasoning.* Baltimore.

Pries, L. (ed.) (2001): *New Transnational Social Spaces. International Migration and Transnational Companies in the Early Twenty-First Century.* London, New York.

— (1998): "Transnationale Soziale Räume", in: Beck, U. (ed.): *Perspektiven der Weltgesellschaft.* Frankfurt/Main, 55–86.

PROLUDRA <Projet de Lutte contre la Désertification dans la Vallée du Dra> (1997): Atelier d'information sur la désertification et sur les activités du PROLUDRA. Zagora (unpublished).

Rachik, H. (2002): "Comment disparaît une norme", in: Mahdi, M. (ed.): *Mutations sociales et réorganisation des espaces steppiques.* Casablanca, 57–64.

— (2000): *Comment rester nomade?* Casablanca.

Rahmani, I. (1998): Description du système de commercialisation ovine dans l'oriental. Cas de la CR de Ain Beni Mathar. Meknès (Mémoire de 3ème cycle en agronomie, ENA, unpublished).

Raikes, Ph./Friis Jensen, M./Ponte, St. (2000): "Global commodity chain analysis and the French filière approach: comparison and critique", in: *Economy and Society* 29,3, 390–417.

Ramdane, A. (1991): Etat des ressources et utilisation des pâturages par les éleveurs nomades: Cas de Missour. Rabat (Mémoire de 3ème cycle, IAV, unpublished).

Rao, A. (2002): "Vanishing Cultures and Struggles for Survival: The crises in Peripatetic Lifestyles", in: *Bulletin of the International Committee on Urgent Anthropological and Ethnological Research* No. 41, 63–80.

Rapley, J. (2004): *Globalization and Inequality: Neoliberalism's Downward Spiral.* Boulder.

Rass, N. (2006): *Policies and Strategies to Address the Vulnerability of Pastoralists in Sub-Saharan Africa.* (PPLPI Working Paper No.37) Rom.

Raynal, R. (1952): "Les transformations du Maroc oriental", in: *Notes marocaines* 1.

RBA <Réserve Biosphère Arganeraie> (ed.) (1997/2001): *Étude du plan cadre da le Réserve Biosphère Arganeraie: Description et évolution de la région de l'Arganeraie.* Agadir, Rabat.

RGPH <Récensement Général de la Population et de l'Habitat> (1971, 1982, 1994, 2004): Data from Maroccan Population Census. Rabat, Direction de la Statistique.

Rogers, B. L./Schlossman, N. P. (eds.) (1990): *Intra-Household Resource Allocation: Issues and Methods for Development Policy and Planning.* Tokyo.

Ruf, U. P. (1995): *Mobil Sesshafte. Sedentarisierung und Geschichte der Nomaden in Mauretanien.* (Bielefelder Studien zur Entwicklungssoziologie 61) Saarbrücken.

Salzman, P. C. (2004): *Pastoralists: Equality, Hierarchy, and the State.* Oxford.

Sassen, S. (1996): *Losing Control? Sovereignty in an Age of Globalization.* New York.

Schlee, G. (2005): "Forms of Pastoralism", in: Leder, St./Streck, B. (eds.): *Shifts and Drifts in Nomad-Sedentary Relations.* (Nomaden und Sesshafte 2) Wiesbaden.

Schmidt, M./Werner, J. (2003): Transition from Mobile Pastoral to Agricultural Land Use in South Morocco – Discussing Changes over Three Decades Using LANDSAT Data. Berlin (unpublished paper).

Schmidt-Wulffen, W. D. (1987): "10 Jahre entwicklungstheoretische Diskussion. Ergebnisse und Perspektiven für die Geographie", in: *Geographische Rundschau* 39, 130–5.

Scholz, F. (1995): "Nomadismus: Theorie und Wandel einer sozio-ökologischen Kulturweise", in: *Erdkundliches Wissen* 118.

Scoones, I. (1998): *Sustainable Rural Livelihoods. A framework for Analysis.* (IDS Working Paper No. 72) Brighton.

— (1996): *Living with Uncertainty. New Directions in Pastoral Development in Africa.* London.

Scott, J. W. (1991): "The Evidence of Experience", in: *Critical Inquiry* 17, 773–97.

Semsek, H.-G. (1987): "Popular Culture versus Mass Culture: The Social Dynamics in a Popular Cairene Quarter. A Case Study", in: Stauth, G./Zubaida, S. (eds.): *Mass Culture, Popular Culture and Social Life in the Middle East.* Stuttgart, Wiesbaden.

Shamir, R. (2005): "Without Borders? Notes on Globalization as a Mobility Regime", in: *Sociological Theory* 23,2, 197–217.

Shelby, D. (2004): "U.S. and Morocco Conclude Free Trade Agreement: Parties See Agreement as a Building Block in Regional Free Trade Area", in: *Washington File* (<http://usinfo.state.gov/>, Access at Mar. 2004), Washington, D.C.

Shiva, V. (1999): "Resources", in: Sachs, W. (ed.): *The Development Dictionary. A Guide to Knowledge as Power.* London, New York, 206–18.

Skounti, A. (1995): Le sang et le sol. Les implications socioculturelles de la sédentarisation. Cas des nomades Ayt Merghad. (unpublished doctoral thesis) Paris.

Slyomovics, S. (2003): "No Buying off the Past: Moroccan Indemnities and the Opposition", in: *Middle East Report* 33, 34–7.

— (2001): "A Truth Commission for Morocco", in: *Middle East Report* 31, 18–21.

Smith, J./Wallerstein, I./Evers, H.-D. (eds.) (1984): *Households and the World-Economy.* Beverly Hills, London.

Spillmann, G. (1936a): "Nomadisme et sédentarisation en pays Ait Atta", in: *Bulletin Economique et Social du Maroc [BESM]* 12, 97–102.

— (1936b): *Les Ait Atta du Sahara et la pacification du Haut Dra.* Rabat.

Spittler, G. (1998): *Hirtenarbeit. Die Welt der Kamelhirten und Ziegenhirtinnen von Timia.* Köln.

Spivak, G. (1988): "Can the Subaltern Speak?" in: Nelson, C./Grossberg, L. (eds.): *Marxism and the Interpretation of Culture.* London, 217–313.

Stammler, F. (2005): *Reindeer Nomads Meet the Market. Culture, Property and Globalisation at the 'End of the Land'.* (Halle Studies in the Anthropology of Eurasia Vol. 6) Halle/Saale.

Steinmann, S. H. (2002): "Gender, Pastoralism, and Intensification: Changing Environmental Resource Use in Morocco", in: *Yale F&ES Bulletin* 103, 81–107.

— (2001): Gender, Pastoralism and Intensification: Changing Patterns of Resource Management in Eastern Morocco. Worcester (unpublished PhD thesis).

Streck, B. (2002): "Systematisierungsansätze aus dem Bereich der ethnologischen Forschung", in: *Orientwissenschaftliche Hefte* 3, *Mitteilungen des SFB „Differenz und Integration" 1: Nomadismus aus der Perspektive der Begrifflichkeit*, 1–10.

Swearingen, W. D. (1996): "Agricultural Reform in North Africa: Economic Necessity and Environmental Dilemmas", in: Vandewalle, D. (ed.): *North Africa: Development and Reform in a Changing Economy.* New York, 67–92.

— (1987): *Moroccan Mirages. Agrarian Dreams and Deceptions 1912–1986.* Princeton.

—/Bencherifa, A. (eds.) (1996): *The North African Environment at Risk*. Boulder.

Swift, J. (1996): "Desertification: Narratives, Winners & Losers", in: Leach, M./Mearns, R. (eds.): *The Lie of the Land: Challenging Received Wisdom on the African Environment.* London, 73–90.

— (1989): "Why are Rural People Vulnerable to Famine?", in: *IDS Bulletin* 20,2, 8–15.

— (1988): *Major Issues in Pastoral Development with Special Emphasis on Selected African Countries* (FAO) Rome.

Swyngedouw, E. (2005): "Dispossing H2O: The Contested Terrain of Water Privatization", in: *Capitalism, Nature, Socialism* 16, 81–98.

Szersynski, B./Urry, J. (2006): "Visuality, Mobility and the Cosmopolitan: Inhabiting the Word from afar", in: *The British Journal of Sociology* 57,1, 113–31.

Tag, B. (2003): *Espace et Société Agro-pastorale en Mutation dans le Maroc Oriental Steppique.* (Série Thèses et Monographies No. 5) Fès.

— (1984): *Les Hauts Plateaux du Maroc Oriental: La destruction du nomadisme et ses conséquences, essai de synthèse.* (No. Spécial de Géographie) Fès.

Thiault, M. (1994): *Essai d'analyse de la situation des parcours de Fezouata.* Ouarzazate.

Thompson, E. P. (1993): *Customs in Common*. New York.

Thornton, P. K./Kruska, R. L./Henninger, N./Kristjanson, P. M./Reid, R. S./Atieno, F./ Odero, A. N./Ndegwa, T. (2002): *Mapping Poverty and Livestock in the Developing World.* Nairobi.

Tietze, B. (1998): "Neue Nomaden – nomadische Arbeitskulturen: Zukunftsprognosen für die kulturelle Entwicklung der Industrie, der Arbeit und des Designs", in: *Straßenfeger*, 8–9.

Tomlinson, J. (1999): *Globalization and Culture.* Oxford.

Touabi, H. (1994): Système d'élevage et pratiques des éleveurs. Cas de la CR de Missour. Meknès (Thèse de 3ème cycle en agronomie, ENA, unpublished).

Tozy, M. (2002): "Des tribus aux coopératives ethno lignagères. Histoire d'une mutation en cours sur les hauts plateaux de l'Oriental", in: Mahdi, M. (ed.): *Mutations sociales et réorganisation des espaces steppiques.* Casablanca, 19–38.

Troin, J.-F. (ed.) (2002): *Maroc. Régions, Pays, Territoires.* Paris.

— (1975): *Les souks marocains: marchés ruraux et organisation de l'espace dans la moitié nord du Maroc.* Aix-en-Provence.

Turner, B. (2005): "Der Wald im Dickicht der Gesetze: Transnationales Recht und lokale Rechtspraxis im Arganwald (Marokko)", in: *Zeitschrift für Entwicklungsethnologie* 14,1, 97–117.

— (2003): "Chr^{c}ka in Southwest Morocco: Forms of Agrarian Cooperation between Khammessat System and Legal Pluralism", in: Pradhan, R. (ed.): *Legal Pluralism and Unofficial Law in Social, Economic and Political Development.* (Papers of the XIIIth international congress, 7–10 April 2002, Chiang Mai, Thailand) Kathmandu, 227–55.

— (2001): "Die Persistenz traditioneller Konfliktregelungsverfahren im Souss, Marokko", in: Fikentscher, W. (ed.): *Begegnung und Konflikt – eine kulturanthropologische Bestandsaufnahme.* München, 187–202.

— (2000): "Sustainable Development and Indigenous Conceptions of Natural Resources Management in the Argan Forest of SW Morocco", in: Castro, M. et al. (eds.): *Actas des XII Congreso International de la Comisión de Derecho Consuetudinario y Pluralismo Legal, Tomo I.* Santiago, 490–504.

Turner, M. D. (2003): "Methodological Reflections on the Use of Remote Sensing and Geographic Information Science in Human Ecological Research", in: *Human Ecology* 31, 255–79.

UNCCD (2004): Report on the side event meeting UNCCD CRIC (Committee to Review Implementation of the Convention) 1. Pastoralists: Custodian of Drylands. Rome, Italy, in: http://www.iucn.org/themes/ceesp/wkg_grp/SL/CRIC-report.pdf. Access at 06 Feb. 2004.

UNDP <United Nations Development Programme> (2005): *Human Development Report 2005.* New York.

USAID <United States Agency for International Development>/ MARA/ USU <Utah State University> (1986): *Final Report: Activities, Findings and Conclusions of the Range Management Improvement Project 608–0145, Morocco.* Logan.

USTR <United States Trade Representative> (2004): *Statement of U.S. Trade Representative Robert Zoellick following House Approval of Morocco Free Trade Agreement.* Washington, D.C.

Urry, J. (2004): "The New Mobility Paradigm", in: Bonß, W./Kesselring, S./Vogl, G. (eds.): "Mobility and the Cosmopolitan Perspective" (Working paper series to Workshop at the Munich Reflexive Modernization Research Centre [SFB 536], 29–30 January 2004, Munich), in: *<http://www.sfb536.mwn.de/veranstaltungen/B3_Workshop_0104_Dokumentation.pdf>* (Access at Sept. 2006).

— (2000a): "Mobile Sociology", in: *British Journal of Sociology* 51,1, 185–203.

— (2000b): *Sociology beyond Societies. Mobilities for the Twenty-First Century.* London, New York.

Vajda, L. (1986): *Untersuchungen zur Geschichte der Hirtenkulturen.* Wiesbaden.

Van Buu, E. (1997): "Chronique juridique Marocaine", in: *Annuaire de l'Afrique du Nord* 36, 687–700.

Van Dijk, J. W. M. (2000): "Régimes fonciers et ménagement des ressources dans un contexte pluriethnique et de pluralisme juridique", in: Diallo, Y./Schlee, G. (eds.): *L'éthnicité peule dans des contextes nouveaux.* Paris, 37–64.

Van Dillen, S. (2004): *Different Choices: Assessing Vulnerability in a South Indian Village.* (Studies in Development Geography) Saarbrücken.

Vialatte, D. (1954): "La situation critique de l'élevage dans le territoire de Ouarzazate", in: *Bulletin Economique et Social du Maroc [BESM]* 61, 47–57.

Waibel, T./Bourhili, A./Miguil, A./Hammoudou, M. (1993): Étude des aménagements sylvo-pastoraux. Ouarzazate (unpublished).

Wallace, C. (2002): "Household Strategies: Their Conceptual Relevance and Analytical Scope in Social Research", in: *Sociology* 36,2, 275–92.

Watts, M. J./Bohle, H.-G. (1993): "The Space of Vulnerability. The Causal Structure of Hunger and Famine", in: *Progress in Human Geography* 17, 43–69.

Weber, M. (1972): *Wirtschaft und Gesellschaft.* Tübingen.

Wellman, B. (2000): "Changing Connectivity: A Future History of Y2.03K", in: *Sociological Research Online* 4,4 (<htpp://www.socresonline.org.uk/4/4/wellman.html>).

Werlen, B. (1997): *Sozialgeographie alltäglicher Regionalisierungen. Globalisierung, Region, Regionalisierung.* Stuttgart.

Werner, J. (2006): *Nomades entre marginalisation, entreprenariat et conflits: Stratégies des éleveurs mobiles du sud marocain faisant face à des bouleversements et engagements requis des acteurs impliqués en vue d'un pastoralisme durable.* Weikersheim.

— (2003): Elevage mobile dans l'arganeraie. Une analyse de la situation actuelle et de sa problématique. Agadir (unpublished report).

— (2002): Conduite économique des ménages des éleveurs mobiles. Cinq études de cas dans la vallée moyenne du Drâa. Zagora (unpublished report).

Wiese, M. (2004): *Health-Vulnerability in a Complex Crisis Situation. Implications for Providing Health Care to Nomadic People in Chad.* (Studies in Development Geography) Saarbrücken.

Williams, R. (1973): *The Country and the City.* Chatto, Windus, London.

Wong, D. (1984): "The Limits of Using the Household as a Unit of Analysis", in: Evers, H.-D./Smith, J./Wallerstein, I. M. (eds.): *Households and the World-Economy.* Beverly Hills, London, 56–63.

World Bank (2006a): *Kingdom of Morocco. Country Economic Memorandum. Fostering Higher Growth and Employment with Productive Diversification and Competitiveness. Volume I: Synthesis.* (Report No. 32948-MOR).

— (2006b): *Kingdom of Morocco. Country Economic Memorandum. Fostering Higher Growth and Employment with Productive Diversification and Competitiveness. Vol. II: Background Studies.* (Report No. 32948-MOR).

— (2005a): *Country Assistance Strategy for the Kingdom of Morocco.* Washington, D.C.

— (2005b): *Morocco at a Glance.* (<www.WorldBank.org/data>, Access at Jan. 2006) Washington, D.C.

— (2005c): *World Development Indicators.* Washington, DC.

— (2004): *Kingdom of Morocco Poverty Report.* Washington, D.C.

— (2003a): *Morocco at a Glance.* (<www.WorldBank.org/data>, Access at Mar. 2004) Washington, D.C.

— (2003b): *Project Appraisal Document on a Proposed Loan to the Kingdom of Morocco for the Rainfed Agriculture Development Project.* Washington, D.C.

— (2001): *The Kingdom of Morocco Country Assistance Strategy, Public Information Notice.* Washington, D.C.

— (1997): *Kingdom of Morocco Country Assistance Review.* Washington, D.C.

— (1995): *Kingdom of Morocco: Water Sector Review.* Washington, D.C.

— (1994a): *Kingdom of Morocco: Poverty, Adjustment & Growth.* Washington, D.C.

— (1994b): *Kingdom of Morocco: Second Agricultural Sector Investment Loan.* Washington, DC.

— (1987): *World Development Report.* New York, Oxford.

Zainabi, A.T. (1989): "Vers une disparition rapide du nomadisme au Sahara Marocain (1): Le cas du Dra Moyen", in: URBAMA (ed.): *Le Nomade, l'Oasis et la Ville.* (Fascicule de Recherche No. 20) Tours, 49-62.

Zougoulou, A.M. (1992): Contribution à la connaissance du pastoralisme nomade dans la commune rurale de Missour. Rabat (Mémoire de 3ème cycle en agronomie, IAV, unpublished).

List of Maps, Figures and Tables

List of Maps

Map 1-1	Case Studies in *Pastoral Morocco*	4
Map 1-2	Morocco – Distribution and Variability of Precipitation	5
Map 1-3	Morocco – Spatial Distribution of Population	6
Map 1-4	Morocco – Rural Poverty	7
Map 3-1	Morocco 1934 – Agricultural and Pastoral Production Systems	34
Map 3-2	Morocco 2005 – Agricultural and Pastoral Production Systems	35
Map 3-3	Morocco – Labor Migration (1934, 2005)	38
Map 5-1	The Urban Corridor of the High Plateaus	89
Map 6-1	The PDPEO Zone – Administrative Setting	96
Map 6-2	Geography of Interventions in the PDPEO Zone	97
Map 9-1	Vegetation of the Collective Rangeland Area Mijamouene	138
Map 9-2	Daily Movement of Sheep Flocks in the Collective Rangeland Area Mijamouene	139
Map 10-1	Massif of Oulmes – Agro-Sylvo-Pastoral Land Use Patterns	152
Map 10-2	Ethno-Territorial Organization of the Ait Amar	153
Map 11-1	Labor Mobility in Oussikis	169
Map 11-2	Pastures Used by Oussikis Pastoralists (2001–2003)	177
Map 12-1	Movement Pattern of One Pastoralist Based in Tiznit (1994–2003)	183

List of Figures

Figure 3-1	Annotations for Map 3-1 and Map 3-2	37
Figure 7-1	Sheep Commodity Chains in Eastern Morocco (Numbers in 1,000 Head)	109
Figure 9-1	Traditional Utilization of Natural Range and Crop Areas by the Ireklaouen	135
Figure 9-2	Seasonal Ewe Productivity Index for Nomadic, Transhumant, and Sedentary Flocks	143
Figure 9-3	Cover of Shrub Species of Two Pasture Types in the Mijamouene Collective Area (in percent)	145
Figure 10-1	Levels of Tribal Structure of the Ait Amar	151
Figure 12-1	Livelihood Strategies of Pastoralists from the Draa Area	185

List of Tables

Table 3-1	The Evolution of Nomadism in Southeast Morocco: Numbers of Nomads	33
Table 3-2	Number of Cattle, Sheep, and Goats in 1932, 1958, and 2004	42
Table 4-1	Development Indicators for the Maghreb (2003–2004)	67
Table 6-1	Eastern Moroccan Highlands – Population	95
Table 6-2	Eastern Moroccan Highlands – Annual Population Growth	95
Table 6-3	Eastern Moroccan Highlands – Internal and International Emigration	99
Table 8-1	Social Stratification in Miqbi Village According to Livestock Production	124
Table 9-1	Classification of Sheep Production Systems by the Ireklaouen in the Middle Atlas	136
Table 9-2	Characterization of Sedentary, Transhumant, and Nomadic Husbandry Systems in the Middle Atlas	141
Table 9-3	Reproductive Performance and Productivity of Ewes According to Husbandry System and Lambing Season	142

Table 9-4	LS-Means and Standard Errors for Effects of Interaction between Pasture Type and Season on Primary Production	146
Table 10-1	Vulnerability Indicators of the Rural Populations in Oulmes	159
Table 11-1	Economic Activities of the Male Adults of Oussikis	167
Table 11-2	Types of Pastoral Households in Oussikis	170
Table 11-3	A Scheme for Quantitative Livelihood Assessment	172
Table 11-4	Pastoral Livelihoods – Resource Access in *Irahhaln* Households	173
Table 11-5	Mobility Patterns and Herd Management of the Ait Oussikis Nomads	176
Table 12-1	Movement Pattern of one Pastoralist Based in Tiznit (1994–2003)	182
Table 14-1	Innovations and their Adoption among the Beni Guil	215

Nomaden und Sesshafte

Sonderforschungbereich Differenz und Integration.
Wechselwirkung zwischen nomadischen und sesshaften Lebensformen in Zivilisationen der Alten Welt
Hg. im Auftrag des SFB von Stefan Leder und Bernhard Streck

Die Reihe befasst sich mit den Beziehungen zwischen Nomaden und Sesshaften in Geschichte und Gegenwart. Untersuchungsraum sind die Trockengürtel Nordafrikas und Asiens, in denen mobile und sesshafte Lebensformen seit jeher in enger Verbindung stehen. Handel und Austausch, Abgrenzung und Konflikt, Versuche der Beherrschung und Prozesse der Durchdringung bestimmen das Verhältnis von festen, oft urbanen Siedlungsformen und nomadischen Bevölkerungen. Diese wechselvolle Koexistenz hat die Zivilisationen in weiten Teilen der Welt über lange Zeiträume geprägt und ist bis heute wirksam.

Band 1: Stamm und Macht

Die arabischen Stämme im 2. und 3. Jahrhundert der Hiğra
Von Eva Orthmann
552 S., 3 s/w-Abb., 2 Karten, geb., (978-3-89500-288-5)

Diese Studie nimmt erstmals Stammeskonflikte aus dem 2. und 3. Jahrhundert der Hiğra systematisch in den Blick. Dabei wird deutlich, dass der Wechsel von den Umaiyaden zu den Abbasiden für die Stämme ein weniger einschneidendes Ereignis dargestellt hat, als bisher vermutet. Unter Hinzuziehung ethnologischer Forschungsergebnisse wird das Verhalten einzelner Clans und Stammesführer analysiert, und insbesondere nach den Eigeninteressen kleiner Stammeseinheiten gefragt. Zunehmend kommen hierdurch Zweifel an der praktischen Bedeutung übergeordneter Stammeskonföderationen auf. Die Studie stellt daher gängige Erklärungsmuster, die von einer polaren Spaltung der Stämme in Qais und Yaman ausgehen, auch für die Umaiyadenzeit in Frage.

Band 2: Shifts and Drifts in Nomad-Sedentary Relations

Ed. by Stefan Leder and Bernhard Streck
516 S., 21 Abb., geb., (978-3-89500-413-1)

The studies contained in this book focus on the impact of interrelations between nomadic and sedentary societies. The authors, anthropologists and historians, have examined a wide range of nomad-sedentary relations and have discussed the effects of these interrelationships. Their inquiry exposes the diversity and flexibility characteristic of nomadic economy, social organization and practices, as they explain how these determine, and result from, interaction with sedentary social environments. The topics include ancient Egypt, North-Africa in Roman antiquity, the Near East from late antiquity till modern times, East-Africa, Iran and Central Asia, as well as gypsy groups in Turkey and in the Black Sea area. This comparative perspective, and also observations concerning the fluidity of boundaries between both ways of life have encouraged the development of a deeper understanding for the systematic aspects of nomadic life. Historical case studies have detected nomad-sedentary relations in several fields, such as military organisations, administration and political institutions. Their analysis correlates historical incidence to circumstantial and recurrent conditions.

Band 3: Die Figur des Beduinen in der arabischen Literatur
Von Sara Binay
264 S., geb., (978-3-89500-531-2)

Der Beduine ist eine Figur, die in allen Textsorten der arabischen Prosaliteratur der so genannten klassischen Epoche verbreitet ist. Sie ist mit einem festen Repertoire an Eigenschaften und Verhaltensweisen ausgestattet. Seine Ausprägung speist sich einerseits aus der Lebenswelt von Nomaden, andererseits resultiert die Spezifik der Beduinenfigur aus ihrer Verwendung in bestimmten Themen des arabischen Diskurses, z. B. Herrschaft, Religiosität, Geschlechterfragen. Die Beduinenfigur der arabischen Literatur, die sich durch ihre Ambivalenz und Ambiguität auszeichnet, eignet sich in besonderem Maße dazu, heikle Themen anzusprechen und Standpunkte in den Diskurs einzubringen, die die geforderte politische, religiöse und soziale Anpassung subtil unterlaufen.
Dieses Buch fügt sich als Beitrag zur Diskursivität in den Kontext von Wechselwirkungen zwischen nomadischen und sesshaften Kulturen ein.

Band 4: Arms and Armour as Indicators of Cultural Transfer
The Steppes and the Ancient World from Hellenistic Times to the Early Middle Ages
Ed. by Jürgen Tubach and Markus Mode
in Cooperation with G. Sophia Vashalomidze
520 S., 219 Abb., 3 Tabellen, geb., (978-3-89500-529-9)

This volume gives an impression of contemporary archaeological, art historical, and historical research in military equipment of and under the influence of nomadic peoples in Eurasian history.
Arms and armour have to be explained within their archaeological context – that is, as parts of a complex system of remains from past societies. Seen from this perspective, they explain, beyond their primary character as military sources, the broader sphere of life of nomadic peoples, their efforts in technology and in art, and key aspects of their spiritual world.

Band 5: Vom Beutezug zur Territorialherrschaft
Das lange Jahrhundert des Aufstiegs von Beduinen zur Vormacht in Syrien und Mesopotamien (286–420/899–1029)
Von Kurt Franz
Ca. 232 S., ca. 4 Abb., geb., (978-3-89500-530-5)

Band 6: Die Alttürkenzeit in Zentralasien
Studien zu Archäologie und Geschichte pastoralnomadischer Gruppen und ihrer Interaktion mit Sesshaften vom 6.–8. Jh.
Von Sören Stark
Ca. 456 S., geb., (978-3-89500-532-9)

Band 3: Die Figur des Beduinen in der arabischen Literatur
Von Sara Binay
264 S., geb., (978-3-89500-531-2)

Der Beduine ist eine Figur, die in allen Textsorten der arabischen Prosaliteratur der so genannten klassischen Epoche verbreitet ist. Sie ist mit einem festen Repertoire an Eigenschaften und Verhaltensweisen ausgestattet. Seine Ausprägung speist sich einerseits aus der Lebenswelt von Nomaden, andererseits resultiert die Spezifik der Beduinenfigur aus ihrer Verwendung in bestimmten Themen des arabischen Diskurses, z. B. Herrschaft, Religiosität, Geschlechterfragen. Die Beduinenfigur der arabischen Literatur, die sich durch ihre Ambivalenz und Ambiguität auszeichnet, eignet sich in besonderem Maße dazu, heikle Themen anzusprechen und Standpunkte in den Diskurs einzubringen, die die geforderte politische, religiöse und soziale Anpassung subtil unterlaufen.
Dieses Buch fügt sich als Beitrag zur Diskursivität in den Kontext von Wechselwirkungen zwischen nomadischen und sesshaften Kulturen ein.

Band 4: Arms and Armour as Indicators of Cultural Transfer
The Steppes and the Ancient World from Hellenistic Times to the Early Middle Ages
Ed. by Jürgen Tubach and Markus Mode
in Cooperation with G. Sophia Vashalomidze
520 S., 219 Abb., 3 Tabellen, geb., (978-3-89500-529-9)

This volume gives an impression of contemporary archaeological, art historical, and historical research in military equipment of and under the influence of nomadic peoples in Eurasian history.
Arms and armour have to be explained within their archaeological context – that is, as parts of a complex system of remains from past societies. Seen from this perspective, they explain, beyond their primary character as military sources, the broader sphere of life of nomadic peoples, their efforts in technology and in art, and key aspects of their spiritual world.

Band 5: Vom Beutezug zur Territorialherrschaft
Das lange Jahrhundert des Aufstiegs von Beduinen zur Vormacht in Syrien und Mesopotamien (286–420/899–1029)
Von Kurt Franz
Ca. 232 S., ca. 4 Abb., geb., (978-3-89500-530-5)

Band 6: Die Alttürkenzeit in Zentralasien
Studien zu Archäologie und Geschichte pastoralnomadischer Gruppen und ihrer Interaktion mit Sesshaften vom 6.–8. Jh.
Von Sören Stark
Ca. 456 S., geb., (978-3-89500-532-9)